Fossils, Paleontology, and Evolution

Fossils, Paleontology, and Evolution

Second Edition

David L. Clark
University of Wisconsin-Madison

ucb
Wm. C. Brown Company Publishers
Dubuque, Iowa

Consulting Editor
Sherwood D. Tuttle
University of Iowa

Copyright © 1968, 1976 by
Wm. C. Brown Company Publishers

Library of Congress Catalog Card Number: 75-21314

ISBN 0-697-05015-7

Printed in the United States of America.

Contents

Preface

This book is an organized review of the sphere of paleobiology. It includes a brief history of paleontology and the origin of life and also a survey of the evolution of the major groups of organisms that have left a fossil record. The final section is a discussion of how fossils are used as tools in the interpretation of Earth history and includes examples ranging from fossil behavioral studies to paleogeophysics. Finally, there is a brief discussion of the possibility of extraterrestrial life, and whether, if such life exists, it would resemble life as we know it on Earth.

Most of the illustrations were prepared by Dorothea Fuchs, Bonn. Judy Meyer, Louise Clark, and Paul Dombrowski, Madison, contributed as well.

David L. Clark

Acknowledgments

The following authors and publishers have granted permission to adapt or use illustrations from their work:

H. N. Andrews, 1961. *Studies in Paleobotany.* New York: John Wiley & Sons, Inc. Used with permission of author and publisher.

A. H. Cheetham, "Late Eocene Zoogeography of the Eastern Gulf Coast Region," *Geological Society of America Memoir 91,* 1963. Used with permission of author and The Geological Society of America.

E. H. Colbert, *Evolution of the Vertebrates,* New York: John Wiley & Sons, Inc., 1955. Used with permission of publisher.

William C. Darrah, *Principles of Paleobotany,* 2d ed., Copyright © 1960, The Ronald Press Company, New York.

William H. Easton, *Invertebrate Paleontology.* Copyright © 1960 by W. H. Easton. Reprinted by permission of Harper & Row, Publishers.

Eldredge, N. and Gould, S. J., "Punctuated equilibria: an alternative to phyletic gradualism," in *Models in Paleobiology,* T. J. M. Schopf, ed. Freeman, Cooper and Co., 1972. Used with permission of publisher.

Fooden, J., "Breakup of Pangaea and isolation of relict mammals in Australia, South America and Madagascar," *Science,* vol. 175, p. 894–898, Fig. 1, 1972. Copyright 1972 by the American Association for the Advancement of Science. Used with permission of author and publisher.

Invertebrate Fossels by R. C. Moore, C. G. Lalicker and A. G. Fischer. By permission of McGraw-Hill Book Company.

H. M. Muir-Wood, and G. A. Cooper, "Morphology, Classification and Life Habits of the Productoidea (Brachiopoda)," *Geological Society of America Memoir 81,* 1960. Used with permission of the authors and the Geological Society of America.

Newell, N. D., "Crises in the history of life," copyright © 1963 by Scientific American, Inc. All rights reserved.

A. S. Romer, *Vertebrate Paleontology* 2nd ed., Chicago: The University of Chicago Press, 1950. Used with permission of author and publisher.

Seilacher, A., "Swimming habits of belemnites-recorded by boring barnacles," *Palaeogeography, Palaeoclimatology, Palaeoecology,* Elsevier Scientific Publisher Co., 1968. Used with permission of author and publisher.

E. C. Stumm, "Silurian and Devonian Corals of the Falls of the Ohio," *Geological Society of America Memoir 93,* 1964. Used with permission of author and The Geological Society of America.

Treatise on Invertebrate Paleontology, courtesy of The Geological Society of America and The University of Kansas Press.

Wells, J. W., "Problems of annual and daily growth-rings in corals," in *Palaeogeophysics,* S. K. Runcorn, ed. London, Academic Press, 1970. Used with permission of author and publisher.

R. P. Wodehouse, *Pollen Grains,* New York: McGraw-Hill Book Company, 1959. Used with permission of publisher.

Chapter One

Paleontology

Paleontology is the study of life of the past (Fig. 1). Living things are the continuation of unbroken chains of ancestors and descendants stretching back 3 billion years or more (see geologic time scale, Fig. 2). The history of these chains of life is part of the study of paleontology. But paleontology is more. It is also a study of fossil organisms that have

Figure 1. A Cretaceous ammonoid cephalopod from Wyoming, a representative of one of the most important fossil invertebrate groups used in the interpretation of Earth History. (Original approximately 12 inches.)

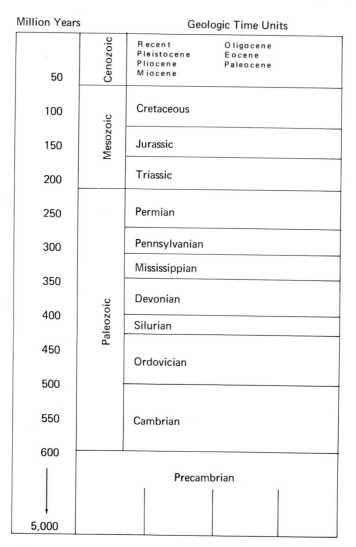

Million Years   Geologic Time Units

Million Years		Geologic Time Units	
50	Cenozoic	Recent Pleistocene Pliocene Miocene	Oligocene Eocene Paleocene
100	Mesozoic	Cretaceous	
150		Jurassic	
200		Triassic	
250	Paleozoic	Permian	
300		Pennsylvanian	
350		Mississippian	
400		Devonian	
450		Silurian	
500		Ordovician	
550		Cambrian	
600			
5,000		Precambrian	

Figure 2. Geologic time scale. (Dates modified from Kulp.)

no living descendants. Extinct forms of life constitute a very significant part of the fossil record. Thus, the unbroken chain that relates living things to ancestors now fossilized, as well as the multitudes of chains that have been broken by extinction, constitute the study of paleontology.

Those groups of fossils that are most fully understood are those having living relatives that can be observed and studied. Fossils representing extinct groups generally are poorly understood because they have no close living relatives that have been identified. The dependency of the scientist upon the relationship between fossil groups and their living representatives in making adequate biologic judgments has been recognized for a long time. This dependency has not always been emphasized, however, and only recently have many students of paleontology concerned themselves with modern life.

Historically, most studies of fossils have been made in connection with applied geological work in which the utilization of fossils to determine the age of the rock in which they occur has been emphasized. This has been done at the expense of understanding these fossils as once-living organisms. In such studies, the identification and relative age of fossil groups have been stressed. This is very important and is still the first step in any systematic study, but we now realize that equally important for the interpretation of Earth history is the recognition that fossils are the remains or at least some evidence of the remains of ancient forms of life. It is important to remember that fossils were living organisms with their own peculiar physical and chemical demands. They required specialized food for their physiologic processes, reproduced under particular environmental conditions, and gave rise to new organisms which, with ancestors and descendants, participated in the continuing process of evolution.

A SHORT HISTORY OF PALEONTOLOGY

As simple as such observations may appear today, it is interesting to note that man has not always understood fossils in this way. In past ages, men have believed many different things concerning fossils, and it has taken hundreds of years for our current ideas to develop. While the advances in the development of modern paleontologic philosophy are at least roughly correlated with the periods of mankind's intellectual development, the achievements of civilization in art, music, and literature in times past, have far surpassed man's understanding of life, its origin, and its development through time.

Men were early concerned with the nature of fossils, but not from the viewpoint of how ancient life aided in the interpretation of Earth History. The first problem was resolving the question: are fossils organic or inorganic? The Greeks were among the first to record their ideas on this subject. The names of Xenophanes, Herodotus, Aristotle, and Theophrastus are prominent among early students, but in this early beginning of the study of natural history, ideas were vague, references to fossils were perfunctory, and the sources for many of these references are questionable.

Evidently, Xenophanes and Herodotus visualized a relationship between fossil sea shells which were found on hillsides and ancient inundations of the land by the sea. However, Herodotus indicated that he, as well as most writers of this time, had little understanding of ancient life when he suggested that certain Eocene fossils which are found in Egypt were the remains of food left by the construction crews of the pyramids.

Concerned more with law, civil matters, and architecture than with natural history, the Romans added little to man's study of ancient life. Lucretius and the Plineys made observations during this period, however, and references by these men and others to currently unknown writings by Greeks and other Romans suggest that more was known, or at least suspected, during these times than has survived to the current generation.

During the late B.C. and early A.D. years, certain vague and mistaken ideas formed the principal philosophy of paleontology. But as small observations were added to meager data, some valid ideas developed and by the time of the Renaissance, men generally believed that fossils were organic in nature, although ideas concerning their origin were still primitive. Men debated whether fossils originated from the fermentation of slime and "fatty matter" in the Earth, were the incomprehensible creations of the Creator, or were made by the Devil to deceive man. In Italy, Leonardo da Vinci, among others, stressed the natural and organic origin and nature of fossils and in the middle of the seventeenth century, Nicolaus Steno of Denmark published his views that molluscs and fossil teeth from "lumps of earth" in Malta were the remains of marine animals and that parts of the island of Malta must, at one time, have been covered by water.

As the fact that fossils were organic in nature and natural in origin became acceptable, different theses were advanced to explain the distribution of certain fossils, obviously marine in origin, on the mountains and hills, places long held to be immutable. During the eighteenth century the Biblical Flood of Noah's time was credited with the mass mortality supposedly evidenced by the fossils of the Earth. Because this was in apparent agreement with religious views, theological leaders gave this idea powerful support for many years. Although this view acknowledged an organic nature and origin for fossils, because it limited the explanation for their existence to a single catastrophic event in history, it did not lead to a systematic philosophy of paleontology.

Paleontology became a science at the beginning of the nineteenth century. Cuvier and Lamarck in France and Smith in England began to lay the foundation of observation which lifted the study of fossils from the doldrums of superstition to a ranking position among the natural sciences. It became apparent to Cuvier, a professor in the College de France in the early 1800s, that the fossil mammals which he pieced together bone by bone were not dead representatives of living species but were extinct types and in many cases were ancestral to living forms. He also noted that fossils in different layers of the Earth generally were of different ages. At approximately the same time, Lamarck, a professor of zoology at the Natural History Museum in Paris, began to accumulate a vast amount of data on invertebrate fossils. His study of animals without backbones led to the famous theory of organic evolution. Lamarck believed that acquired characteristics could be inherited by following generations of organisms and that change of characteristics through time was very much dependent on the environment. This was one of the first serious attempts at a synthesis of ideas on the origin, nature, and distribution of life. Although modified in scope, some of Lamarck's ideas are still considered part of the working hypothesis of organic evolution.

Of equal importance in the elevation of paleontology to a first class science was the work of William Smith in England. Smith, who, in the early nineteenth century, worked on various engineering and surveying teams in extremely fossiliferous strata in England, began to note that certain types of fossils were always associated with specific rock layers. More observations confirmed this and he published several volumes describing his finds, including the first geologic map of England.

In almost every kind of human endeavor, at least early in its history, there must be some practical justification for study, a justification which serves as the catalyst to get other people interested. As more people become interested, more work is done and knowledge increases quickly. Smith demonstrated that fossils were useful and had definite value for the recognition and definition of the layers of the Earth. He was able to show that rock layers separated geographically could be correlated by their contained fossils. Thus, stratigraphic paleontology or biostratigraphy was born. Since that time, the utilization of fossils in the interpretation of rock strata has been a primary motivation for their study.

After Smith, paleontology had a purpose in addition to study for study's sake. Then in 1859, Darwin published his theory of organic evolution and it was recognized that fossils were the primary evidence for this theory. Sequences of fossils from different layers of the Earth were then studied with ancestors and descendants in mind and when such relationships were determined some of the tempo and some of the mechanics of evolution became known for the first time. The reasons for the distribution of fossils and living organisms in time and space could now be explained quite logically.

The study of fossils has branched in many directions during the last hundred years. The value of fossils in the interpretation of the age and relationships of the rock layers of the Earth is universally recognized and studies related to this are still among the most important. In addition, many of the principal lineages or phylogenies have been determined and with the help of Darwin's theory and an understanding of modern genetics, details of the "how" of evolution through millions of years of the history of the Earth are becoming clear. Still, there are many undescribed or "new" species of animals and plants which are yearly culled from the rocks of the Earth. And with each new find, the paleontologist who prepares a careful description of his discovery adds another page to the book of the life and history of the Earth.

Paleontological research moves through a series of steps each designed to give additional knowledge. The descriptive stage is first and then the determination of a fossil's position in time (biostratigraphy) and space (paleozoogeography). Finally, the relationship of a fossil and the group or population to which it belongs, as well as its contemporary life and the dependency between the organism and the chemical and physical factors of its former environment must be determined (paleoecology).

So, the study of paleontology, initiated in the midst of ignorance of natural processes and nurtured by medieval superstition, has emerged in the twentieth century as the champion of evolutionary theory and as a skeleton key to the interpretation of the sedimentary rocks of the Earth.

REFERENCES

Adams, F. D. 1938. *The birth and development of the geological sciences.* New York: Dover (Paperback, Dover, 1954).

Darwin, Charles. 1859. *On the origin of species by means of natural selection, or the preservation of favoured races in the struggle for life.* London: Murray (Paperback, Mentor, 1958).

Dott, R. H., Jr., and Batten, R. L. 1971. *Evolution of the Earth.* New York: McGraw-Hill.

Geikie, A. 1905. *The founders of geology.* 2d ed. New York: Macmillan. (Paperback, Dover, 1962).

Kulp, J. L. 1960. *The geological time scale.* International Geological Congress, Report of the Twenty-First Session, Norden, Part III, pp. 18–27.

Zittel, K. A. von. 1901. *History of geology and paleontology to the end of the nineteenth century.* London: Walter Scott.

Chapter Two

Appearance of Life on Earth

THEORY

Ironically, man's recent trips to the Moon have been motivated, in part, by the hope that information obtained could enlarge our knowledge of the Earth. When did the Earth form? When did life appear on the Earth? Could the rock record of the Moon—or elsewhere—hold clues not found on the Earth?

Study of the Moon has given relatively few surprises but it has made man more secure in his previous knowledge. Our Earth formed as part of the solar system some 4 to 5 billion years ago. At that time no life was present even though all of the factors necessary for primitive life did exist. Biochemists have demonstrated on many occasions that the important prerequisites for the formation of life are ammonia (NH_3), hydrogen (H_2), nitrogen (N_2), water (H_2O), and possibly methane (CH_4) and carbon dioxide (CO_2). The table below shows a list of the chemical elements contained in the atmosphere of the primitive Earth as proposed by students of its origin and early history.

Composition of the primitive atmosphere*

A	B
CH_4	CO_2
NH_3	CO
H_2	H_2
H_2O vapor	N_2
	H_2O vapor

*From Rasool, S. I. 1967. Evolution of the Earth's Atmosphere. *Science* 157:1466–67. According to Rasool, A corresponds to atmospheric composition proposed by Oparin, Miller, Urey, and Ponnamperuma; B to that proposed by Abelson and others.

These chemicals, in the presence of a suitable energy source and in an environment containing little or no free oxygen, can, under controlled laboratory conditions, produce complex organic molecules similar to the basic building blocks of living matter. There are several schools of thought on the exact composition of the materials involved and on the possible sources of energy. Theories differ because of the different scientific backgrounds of those theorizing. Nonetheless, because the production of organic molecules, which are the basic building blocks of living matter, can be obtained from an inorganic system in the laboratory, the real question is whether these conditions ever existed naturally on Earth.

The answer to this important question seems to be positive on all counts. Consider the following:

1. Most of the best theories for the origin of the Earth demand the presence of the very things which the biochemists say would have been necessary for the origin of life. The evidence is based on calculations of the primary chemical constituents of the Earth as well as upon present astronomical observations on the abundance of chemical elements in our solar system. All of these observations seem to indicate that the hydrogen, nitrogen, water vapor, and other chemical items necessary for the origin of life are precisely the compounds which would be present in any primitive atmosphere of a planet such as the Earth.

2. The energy source, whether high energy ultraviolet radiation or a spark of lightning, also would have been available. Recent data from artificial satellites indicate the presence above the Earth's surface of various layers of radiation "belts" or "shields" which must have evolved with our Earth. The absence of these shields early in the Earth's history would have allowed a much greater amount of high energy radiation to come to Earth than can currently pass through the atmosphere. This radiation source was very probably of sufficient magnitude for the actions and reactions necessary for the origin of life.

3. Finally, the last of the three prerequisites for the origin of life is the presence of an oxygen-free or, as it is called chemically, a "reducing" atmosphere. The demonstration that such a condition existed 3 to 4 billion years ago would appear at first glance to be difficult, but geochemists, biologists, and paleoclimatologists have gathered unique data related to this point. First, it is interesting to note that the best theories for the origin of the Earth indicate that oxygen is one of the elements which could not have existed in a free condition when the Earth was formed. Free oxygen was evidently limited to 0.1 percent of the present atmospheric level in the beginning. Fur-

ther, the presence of free oxygen in today's atmosphere is almost completely biologically produced and therefore must have been essentially absent before there was life on the Earth. These biologically deduced facts are significant, but perhaps of equal importance are the results of the studies of the Earth's sedimentary deposits which paleoclimatologists and geochemists have made. Some of these studies were concerned with "oxidized sediment" or, more particularly, the presence of iron-rich sediment which has been exposed to the atmosphere. Oxidation, or rusting, is, of course, common today and most things in our present atmosphere are attacked by oxygen. Oxidized sediment is not present, however, in the rock record prior to 1 billion years ago. The appearance of numerous oxidized sediments or "red beds" in the geologic record evidently coincides with an abundance of oxygen-producing organisms. Before this time, neither enough oxygen nor enough organisms to produce it were present. These observations suggest again, that when the Earth was very young, a reducing atmosphere was present.

Substantiation of the actual presence of the materials, energy, and environment necessary for the production of life is only the first step. More difficult to substantiate are the steps between the formation of life's building blocks and their development into the complex molecules which gave rise to living things. This has been referred to as "molecular ordering," and while actual life production in the laboratory may not be possible at the present time, this molecular ordering does not seem impossible under the proper conditions. Biochemists have demonstrated that inherent electrical charges in molecules may cause the alignment and organization of matter into biologic systems. Molecular ordering has been demonstrated under controlled laboratory conditions which, in turn, can be related to the conditions of the early Earth. All of these factors point to the primeval sea as the setting for the origin of life.

In this primitive and quite sterile environment, inorganic compounds were transformed into organic structures and then molecular ordering proceeded. It was a slow development, but in an environment devoid of free oxygen and existing life forms which might devour them, primitive life-like cells began to form. It has been estimated that a concentration of up to 1 percent of organic material in the sea may have been produced in this way. With time and change and more transformation, primitive life formed; all of the essentials were present and all of the time needed for the slow process was available.

These ideas, theories, and facts give a coherent, if less than completely satisfactory, account of the origin of life. Significant in all observations is the fact that many different scientific disciplines have added

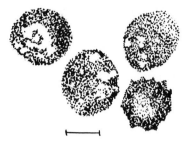

Figure 3. Microscopic cells thought to be the oldest life. Fossils are from the Early Precambrian (3.1 billion years old) of Africa. Bar equals 10 microns. (Adapted from J. W. Schopf)

a fact here, an idea there, and that there is general agreement. Such consistency could be regarded as an indication of the validity of the idea.

FACTS

From theory, we can proceed to factual information concerning the earliest life. Fossils of the oldest and simplest (procaryotic) life forms occur in rocks about 3.1 billion years old (Fig. 3). These most primitive fossils represent forms of bacteria and blue-green algae, descendants of which still are abundant on Earth. More advanced (eucaryotic) life forms had evolved by 1.7 billion years ago and are rather well known because of studies in North America, Australia, and Africa (Fig. 4). Evolution between 3.1 and 1.7 billion years ago has been described as lethargic, but with the development of the nucleated cell about 1.7 billion years ago, the biologic potential for multicellular organisms was present.

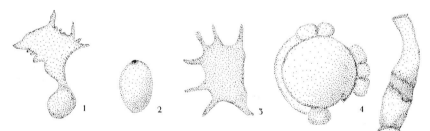

Figure 4. Ancient life. Unusual rod and globular bodies from Precambrian of Canada here magnified approximately 375 times. (Adapted from Barghoorn and Tyler.)

This primitive life can be recognized on the basis of distinctive morphology. In addition, "chemical fossils" consisting of amino acids and other organic substances produced by the organisms have been reported from rocks at least 1 billion years old.

Preservation of this primitive life rarely is satisfactory because hard parts such as shells, bones, and teeth capable of being preserved were not present. Fossils commonly are preserved as *body fossils* (Fig. 5), whole or parts of an organism either preserved in an original condition or preserved in a form in which the original body material has been chemically replaced or altered. Another group of fossils is referred to as *trace fossils* (Fig. 6). These are traces, impressions, tracks, or trails formed by the bodies of organisms. Because trace fossils often are indicative of behavior, these fossils are important to understanding the relationships of fossils to their environment. *Chemical fossils* are the organic residues of life (amino acids, etc.). All three of these kinds of fossils have been found in the older rocks of the Earth.

Early forms of life prospered, reproduced, diversified, and adapted. While many kinds became extinct, the more successful forms left descendants which, in turn, found their niches on Earth and reproduced their kind. Thus life continued. The slow pattern of change is distinct.

The processes involved in change through time are only today being described in the detail necessary for full comprehension, but the fundamental pattern and scope of organic change was first described more than 100 years ago by Charles Darwin. From the origin of life, our observations must now move to its evolution.

EVOLUTION OF LIFE

The Record of Change

Darwin was not the first naturalist to gain an insight into the process of evolution, but his geological background and biological interests endowed him with better equipment with which to build a theory of organic evolution than any of his predecessors. Darwin pointed out that the interaction of a changing environment and a changeable organism would produce varied life forms, each of which was limited by biological makeup to a relatively narrow environmental niche. He observed that changes in characteristics were very slow and that although the ability to make these changes came from "within," the successfulness of any change was related to externally produced conditions. The study of genetics and biochemistry has provided the explanation of the change which comes from "within" and the working of the gene. The study of the structure of the chromosome, deoxyribonucleic acid (DNA) and how mutations may occur, are all part of many secondary school curricula. Since Darwin's time, modern paleontology has been based on

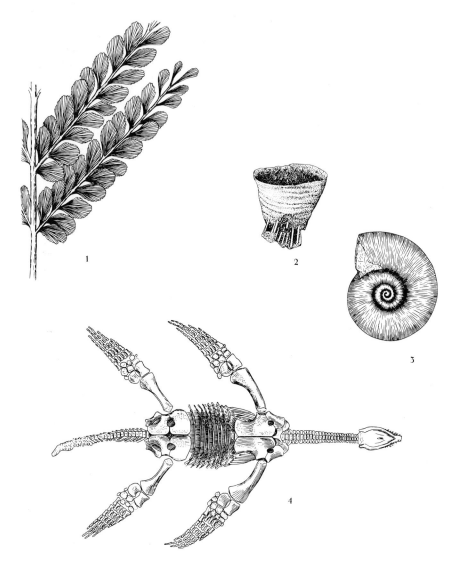

Figure 5. Body fossils. 1. *Archaeopteris hibernica,* an Upper Devonian plant a few inches in length; 2. *Prorichthofenia permiana,* an unusual Permian brachiopod (X 2); 3. *Metalytoceras triboleti,* a Cretaceous cephalopod (X 1/2); 4. *Thaumatosaurus,* a Jurassic plesiosaur, a vertebrate fossil approximately 11 feet long. Invertebrates and vertebrates generally yield the best fossils because they had hard parts capable of being preserved and many lived in an environment (marine water) where sediment burial helped protect the hard parts. (1. adapted from Andrews; 2. after Muir-Wood and Cooper; 3. after Arkell et al.; 4. after Romer.)

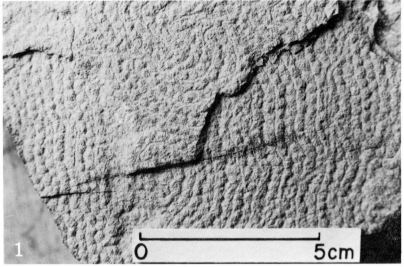

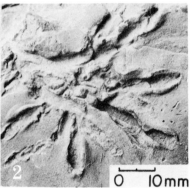

Figure 6. Trace fossils. 1. *Nereites* sp., Pennsylvanian of Oklahoma—meandering feeding trail illustrating intensive working of the sediment (and deep water); 2. *Asterichnus* sp., traces of an animal that worked primarily at the sand-mud interface and pressed sand and mud down below the interface as the radial gallery system developed. (Photographs courtesy C. K. Chamberlain.)

the concept that the fossil record is the firm evidence that such changes have occurred, and have occurred over enormous periods of time.

Genetics and the Reason for Change

One of the paleontologist's tasks is the discovery, study, and interpretation of the record of evolution. While the product of this research provides the facts of evolution, the mechanism of change can be understood best from living organisms. The geneticist and biochemist have given us much information during the last fifty years. The study of genetics has provided the answer to Darwin's problem of the factors "within" which were responsible for the biologic change in characteristics, and the biochemists have explained how the characteristics of organisms are actually produced through the joint action of genes and their complex chemical-physical environment.

We are told that a gene is a complex chemical package of "potential characteristics" and that the number of limbs, the size of bones, or the shape of the shell that an organism may have is dependent upon the action and reaction of a particular set of genes with its environment. Genes are present in every cell and during growth, genes may relay chemical messages and stimulate particular patterns of development. We can, therefore, think of any structure as having resulted from the reaction of one or more genes with its chemical environment.

The genes themselves, present in the germ cell at the time of fertilization, are of tremendous variety, and different environmental stimuli may cause them to relay different messages for different responses in cells. One of the reasons there may be a great number of differences between ancestors and descendants is that there are a large number of genes in any given population, and in sexually reproducing populations, this large number of genes is constantly being reshuffled. One geneticist has pointed out that there are thousands of genes in the human population and from this number of genes there are tens of thousands of possible combinations, each capable of imparting a slightly different characteristic to the recipient. With this number of combinations, the variety of types, even in the human population, may be thought of as surprisingly few. This seems to be related to the fact that in a particular environment there are only a few combinations of characteristics which are really adaptable and while two heads and unchambered hearts and various other kinds of unusual characteristics are possible and appear from time to time, only a single broad type survives in our human population. Any population, whether of clams or elephants, is now visualized as possessing a gene pool from which all of the characteristics of a population—advantageous and harmful—actually are drawn.

This variety of characteristics is relatively unimportant when compared with the results of the change or "mutation" of genes. The fact that a gene may be chemically or physically altered and become a new characteristic potential adds considerable new variety to life. Such changes or mutations occur for a variety of reasons, but the important thing is that they do occur and that this is a continuing process. Because most populations are more or less adjusted to their environments, the majority of gene changes must be considered nonadvantageous. A random, arbitrary change in the mechanism of a clock or some complicated piece of machinery is unlikely to improve its performance. In the same way, because gene mutations are random, they also are usually not advantageous to the organism. Gene mutations occur continuously, however, and this is the primary source of the evolutionary process. The rate at which mutations occur differs in different organisms and may be increased by chemical or physical means. Also, there appear to have been times when evolution occurred much more rapidly than at other times. Many investigators have noticed that the greatest diversification of a group of organisms occurred soon after its initial appearance. Other students have related these times of rapid evolution to changes in the Earth's atmosphere as it journeys through space. Time seems to be the important element in evolution. During the billions of years of Earth history, no external environment has remained the same for great lengths of time. The story of the Earth, as well as that of life, is one of constant change. Where mountains are today, there once was a sea; deserts of the past are now covered by ice; and volcanic ash now covers areas which once were green with lush vegetation.

Each of these environmental changes, and there have been several billion years of them, has challenged the organisms which are dependent upon the temperature, food supply and the many other physical, chemical, and biological factors which comprise an environment. With the possible exception of man, a change in an organism's environment has meant there must be some fundamental change in habit, structure, or function. Evolution takes place on the group level, not on the individual level. Whether or not a group successfully adapts to a change and thus survives is evidently dependent on whether or not there is present in the population a form or forms capable of this kind of adaptation. The ability to adapt must be related to a characteristic or series of characteristics which the particular organism possesses. The appearance of such adaptiveness is related to the number of possible variants introduced by the mutation of genes. If, in any population challenged by an environmental change, there is present a variant which can successfully adapt for any reason, then that population, through the spread of the advantageous character, may itself successfully adapt and survive. In one way of speaking, that population will survive which has the greatest number of useful variants—present and potential. This fact is related, in part, to the number of offspring produced. There is, however, apparently an optimum population size in which advantageous characteristics can arise and spread. Reproduction rate alone was evidently no assurance of the ultimate success or survival of life in the past.

The Mechanism of Change

Visualize, for a general idea of evolution, an environment which is undergoing slow change. This change may take a million or more years to be completed. The rocks of the Earth are a strong testimony to the fact that such changes have occurred. Visualize a population of organisms in this environment, a population which is more or less adapted to the conditions at some point during the time of the change. With this working base, the stage is set for either evolution or extinction. What happens? Certain of the organisms may migrate geographically to areas where "pre-change" conditions can be found, or there may be a group consisting of members which can harmonize with the changing envi-

ronment. If migration or adaptation is not possible (or if the environmental change is too rapid), mass mortality may result.

Whether the challenge thus posed by an environmental modification will cause mass mortality, migration, or encourage adaptation, is
almost entirely dependent upon the population's "change potential" or
gene pool. With the constant reshuffling of genes and the mutations of
others, some variant characteristic may appear, independent of the
reason for the environmental change, but advantageous because of it.
If it does, the advantage imparted to the organism in the face of the
environmental challenge, though small, may be significant. If the advantageous characteristic is disseminated, those individuals properly
endowed will be the favored generation. With time and change, and
time and reproduction, the new but advantageous characteristic may
become common.

The classical picture of evolution has been called "phyletic gradualism." This is descriptive of the slow, progressive process of evolution.
Viewed in this classical manner, phyletic gradualism implies that an
unbroken sequence of ancestors and descendants, each different from
the other by only minor changes in their characteristics, should be
found together in the fossil record. If species A gave rise to species B,
and B to C, etc., then the fossil record should reflect the following
sequence:

$$\text{Species A} \longrightarrow B \longrightarrow C \longrightarrow D \longrightarrow E \longrightarrow F \longrightarrow \text{etc.}$$

Paleontologists have accepted this as their working model for several
generations. There is a problem, however. In most of the well-docu-
mented fossil sequences that have been studied, while ancestors and
descendants may be found in sequence, there are often many missing
links. Thus:

$$\text{Species A} \longrightarrow D \longrightarrow M \longrightarrow R \longrightarrow \text{etc.}$$

This sequence more adequately describes the fossil record than does
the model based on classical Darwinian phyletic gradualism. What the
paleontologist most commonly encounters is more adequately described as "allopatric speciation," a term which describes the idea of the
discontinuous nature of many fossil evolutionary lines. The assumption
for this idea is that the discontinuous nature of the fossil record was
caused by the geographic separation of evolving lineages (Fig. 7). Evolution according to this idea occurs on the fringes of the population more
rapidly than at its geographic center.

Members of a population living in slightly different areas may re-
spond to slightly different environmental conditions in slightly different
manners. Thus genetic changes may be more rapidly spread in geo-

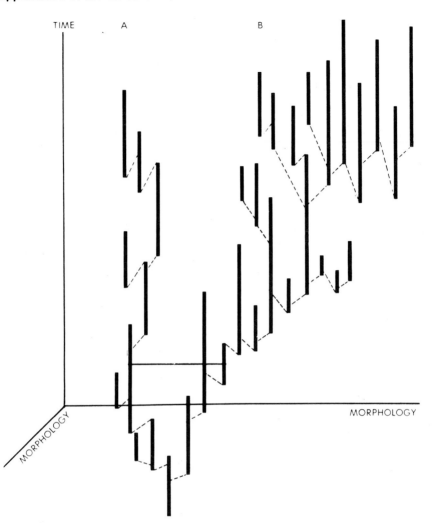

Figure 7. Evolution of species in two lineages, A, and B. This three-dimensional sketch depicts morphological change along the horizontal axes and is a good representation of allopatric speciation. (From Eldredge and Gould.)

graphically separated areas and, thus, evolution from species A,B,C may occur in an area peripheral to the center of activity of the population center. The result may be similar to that illustrated in Figure 7.

Therefore, the sediment record has apparent gaps or missing links and species C apparently succeeds species A, rather than species B. It follows, that a complete geographic sampling, probably impossible, would reveal phyletic gradualism, but the actual record is most likely to represent allopatric speciation. The latter model certainly more closely resembles the fossil record than the traditional model.

EXTINCTION

The record of life is not only one of evolution but also one of extinction. Millions of species of animals and plants have lived and then, unable to adapt to some change in their environment, have become extinct. The familiar stories of the passenger pigeon, the whooping crane, the American buffalo (almost), and the rhinoceros (soon), are those of the alternate to evolution—extinction. The fact of evolution itself involves extinction of a different sort because the very success of adaptation has usually led to new generations whose structure ultimately separates them from their ancestors which then become extinct types.

Estimates as to the total number of species that may have lived range from a few million to as many as 500 million. The fact that there are approximately 1 million species living today (each of which had an ancestor) from among 500 million or so that may have lived, suggests that extinction may be more common than evolution. In addition, we know that there have been times when extinction was more pronounced. That is, periodically great numbers of species became extinct in a relatively short period of time. Such times of great extinction include the Permo-Triassic interval and the Cretaceous-Cenozoic time transition (Fig. 8). Indeed, the recognition and subsequent establishment of the major and minor intervals of geologic time were based on dichotomies caused by extinction. Thus, the Paleozoic Era "ended" and the Mesozoic Era "began," not because this was a random selection of time intervals by some geologist, but because the "end" of the time interval recognized as the Paleozoic is the time during which a major extinction of invertebrates occurred.

A certain periodicity of such extinctions has been noted. Explanations have been proposed that call upon cyclic extraterrestrial events, such as increased cosmic radiation, or galactic changes that might affect

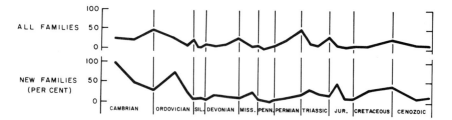

Figure 8. Times of extinction of animals. Top line indicates percent of families of fossil organisms that became extinct. Bottom line indicates number of new families that appeared during indicated intervals. Times of extinction normally are followed by increased evolution and appearance of new groups. (From Newell, N. D. "Crises in the history of life." Copyright © by Scientific American, Inc. All rights reserved.)

a wide range of living things. Periodic reversals of the Earth's magnetic field are known to have occurred during the geologic past. Theoretical calculations suggest that it might be possible that increased cosmic radiation during such times of weakened magnetic field strength (magnetic reversals) may have affected certain organisms. One group of marine plankton evidently shows correlation between times of extinction and times of magnetic field reversals (Hays, 1971). Other groups appear to have been unaffected, and the importance of magnetic field reversals or galactic events on the evolution of life appears questionable.

Although times of extinction are well known, their causes remain a mystery. Recent theories on "continental drift," which involve constant major movements in the Earth's crust, include speculations on what effect these massive migrations of the Earth's surface might have had on these periods of extinction. One suggestion has been made that times of adjustments of various crustal blocks may have so altered the size of continental shelf areas (a principal home of marine invertebrates) that extinctions were produced. So, the periodicity and magnitude of the extinctions remain a matter of considerable interest.

REFERENCES

Barghoorn, E. S., and Tyler, S. A. 1965. Microorganisms from the Gunflint Chert. *Science* 147:563–77.

Dayhoff, M. O. 1972. Evolution of proteins. In *Exobiology,* ed. C. Ponnamperuma. Amsterdam: North Holland Publishers.

Dobzhansky, T. G. 1970. *Genetics of the evolutionary process.* New York: Columbia University Press.

Eldredge, N., and Gould, S. J. 1972. Punctuated equilibria: an alternative to phyletic gradualism. In *Models in paleobiology,* ed. T. J. M. Schopf. San Francisco: Freeman, Cooper.

Hare, P. E., and Abelson, P. H. 1965. Amino acid composition of some calcified proteins. *Carnegie Institution Yearbook, 64.*

Hays, J. D. 1971. Faunal extinctions and reversals of the earth's magnetic field. *Geological Society America Bulletin* 82:2433–47.

Oparin, A. I. 1972. The appearance of life in the universe. In *Exobiology,* ed. C. Ponnamperuma. Amsterdam: North Holland Publishers.

Rhodes, F. H. T. 1967. Permo-Triassic extinction. In *The fossil record,* eds. W. B. Harland, et al. London: Geological Society of London.

Schopf, J. W. 1972. Precambrian paleobiology. In *Exobiology,* ed. C. Ponnamperuma. Amsterdam: North Holland Publishers.

Schopf, T. J. M., ed. 1972. *Models in paleobiology.* San Francisco: Freeman, Cooper.

Simpson, G. G. 1953. *Major features of evolution.* New York: Columbia University Press.

Volpe, E. P. 1967. *Understanding evolution.* Dubuque: Wm. C. Brown Co. Publishers (paperback).

Chapter Three

Fossil Plants

We now have reviewed the origin and evolution of life. Next is a review of the major kinds of fossils and their occurrence in the rocks of the Earth.

As stated earlier, the earliest fossils, primitive bacteria and algae, occur in rocks 3.1 billion years old. From that time to the present, plants of various types have thrived in the Earth's waters, and approximately 400 million years ago, the first land plants appeared. Modern types have evolved in the past 100 million years, but representatives of most of the primitive plants still live.

Although the record of fossil plants does not appear to be as extensive as that of animals, a considerable body of knowledge has accumulated concerning the evolution of plants.

CLASSIFICATION OF PLANTS

The general scheme of evolution within the plant group is fairly well understood. There are fifteen or more major groups of plants which have lived and their classification is rather complex. A simplified outline shows at least four basic groups (not formal taxonomic categories): (1) The primitive types, including algae, fungi, and for convenience, mosses and liverworts; (2) the pteridophytes, including the psilophytes, ferns, lycopods, and horsetails; (3) the gymnosperms, including the seed ferns, cycads, ginkgos, cordaites and conifers; and (4) the angiosperms, which include all of the flowering plants (Fig. 9).

PRIMITIVE PLANTS

The algae, fungi, mosses, and a few other groups in which woody tissue is reduced or absent are generally considered the most primitive

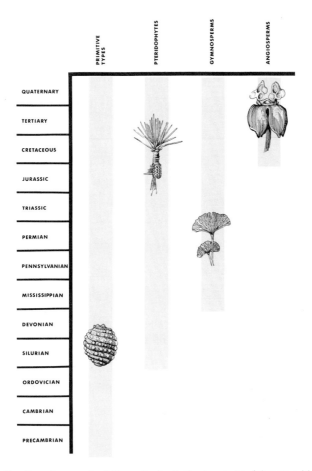

Figure 9. Development of the principal plant groups (shown with some living representatives) and their distribution through time.

types of plants. No one is certain what relationship exists between the oldest life forms which were recently recognized in Canada and South Africa (Fig. 3), and the primitive algal-like fossil structures which have been known for generations.

PTERIDOPHYTES —FIRST VASCULAR PLANTS

Next highest on the evolutionary ladder of plants are the first vascular types—plants which had some form of woody tissue. The oldest definite vascular types are those described from Silurian rocks. Recently, unusual Cambrian plants have been discovered in India, Kashmir, and Russia, which may be closely related to the Silurian types.

Primitive vascular types consist of wood, a central pith, and, in addition, usually two thin layers of different cells which are external to the wood. Early types include forms called psilophytes and are illustrated by *Rhynia,* a dichotomously branching form which stood from a few inches high to a foot or more with no true root system and branches without true leaves (Fig. 10.1–3). These early types reproduced by means of spores—microscopic bodies which were produced in great numbers by sporangia at the top of the branches.

Evolution of these plant forms was rapid and by the Middle Devonian, the first forests with large trees developed. These forests consisted, in addition to *Rhynia* types, of ferns, horsetails, and lycopods or club mosses (Fig. 10.4).

Some of the Carboniferous ferns left magnificent fossils. Impressions and compressions of stems, leaves, roots, and complete plants, 25–50 feet or more in height, have been found (Figs. 11 and 12).

The lycopods or club mosses, including *Lepidodendron,* the so-called scale tree (Fig. 13), stood 100 feet or more, had thick trunks, and are distinguished by an extensive underground root system, and more specifically by a curious pattern of leaf scars on the trunk. The leaves were attached along the length of their base and when they were shed, left prominent scars (Fig. 14). The spiral arrangement of these scars gave a peculiar pattern to the trunks.

GYMNOSPERMS

A varied group of nonflowering plants still abundant today are the gymnosperms. This group includes several types which may not be closely related, but it is convenient to include the seed ferns, cycads, ginkgos, cordaites, conifers, and gnetophytes in the same category.

The seed ferns are a group that shows advancement over more primitive fern types (Fig. 15). The conifers, including pines, firs, spruce, redwoods, and others, include more than 500 species, approximately 80 percent of all living gymnosperms. Several families were abundant from the Pennsylvanian to the Recent. The Arizona Petrified Forest of the Triassic age is primarily a conifer forest and 42 species of conifers and other types have been identified there. The conifers are one of the most successful groups of plants.

The ginkgo has often been described as a "living fossil." This unusual plant (Fig. 16) was abundant in many parts of the world during the late Paleozoic, the Mesozoic, and the early Cenozoic. It became extinct in North America during the Miocene and in Europe during the Pliocene. Some nineteen genera were known before this extinction. In the early 1700s, ginkgos were discovered growing in eastern China where they had been cultivated for centuries in Chinese gardens. The

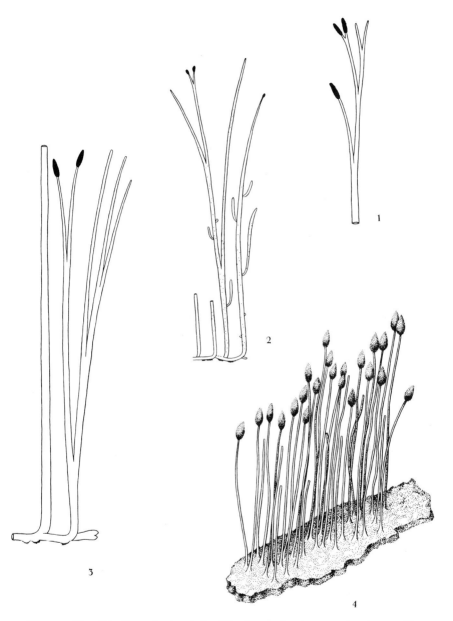

Figure 10. Primitive plants. 1–3. *Rhynia,* a simple vascular type with branching stalks which bear single reproductive bodies. Approximately 1 foot high (after Darrah); 4. restoration of a primitive moss, *Sporogonites exuberans,* approximately 6 inches long, from the Devonian of Belgium. This primitive form shows unbranched stalks and had no vascular tissue (after Andrews).

Figure 11. Reconstruction of a Pennsylvanian fern tree. Below the fronds can be seen leaf scars on this 25 foot tree. Most of the trunk is encased in adventitious roots which thicken toward the base. (After Andrews.)

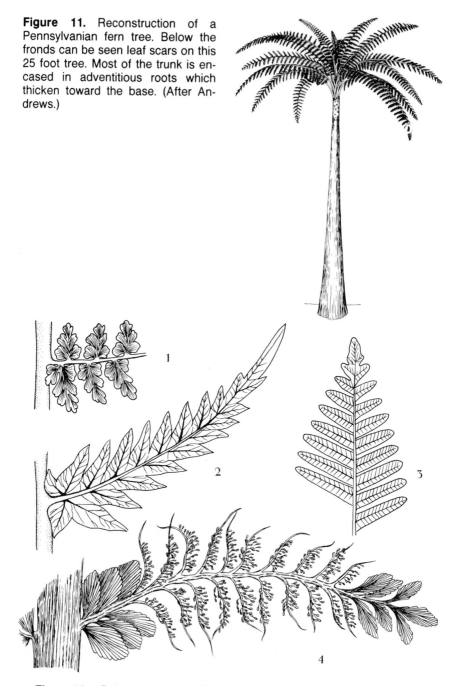

Figure 12. Paleozoic ferns. 1. *Sphenopteris;* 2. *Mariopteris;* 3. *Pecopteris;* all Pennsylvanian ferns approximately natural size (X 1) (after Darrah); 4. *Archaeopteris hibernica,* an Upper Devonian primitive fern with fertile pinnules in middle part (after Andrews).

Figure 13. Restorations of *Lepidodendron,* the scale tree, a Pennsylvanian lycopod; young tree before branching (on right) and older one (on left) approximately 100 feet high. (After Andrews.)

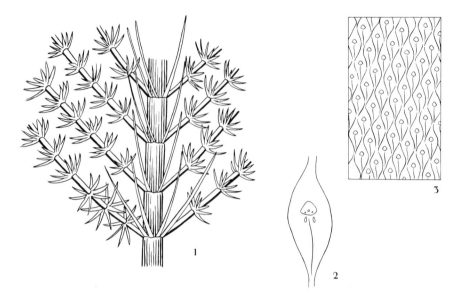

Figure 14. 1. *Annularia,* a common Pennsylvanian calamitid type (approximately X 1/4) (after Darrah); 2., 3. *Lepidodendron* branch exterior (approximately X 1/2). Notice spiral pattern and details in 3. which shows scars enlarged. (Adapted from Andrews.)

plant was transplanted to Europe and then back to North America and now is becoming a popular shade tree.

ANGIOSPERMS

The flowering plants, or angiosperms, appeared during the Mesozoic and are a dominant part of late Mesozoic and Cenozoic floras. The evolution of angiosperms was so rapid that Late Cretaceous floras are more similar to those of the Recent than to those of the Early Cretaceous. This group contains common dicotyledon characteristics and most of the plants are familiar to everyone—oaks, maples, walnuts, poplars, birch, hickory, sycamores, magnolias, ivy, etc. Figure 17 illustrates a typical type. In addition to the dicotyledon types, there are the monocotyledons, such as grasses, and cereals.

An Upper Cretaceous flora from Alaska consisting of more than 200 species included 15 genera of gymnosperms but 73 genera of angiosperms. The evolution of the group shows an enormous diversification and radiation of species. Angiosperms have been the dominant part of most floras since their appearance.

Flowering types reproduce by means of pollen, tiny germinating

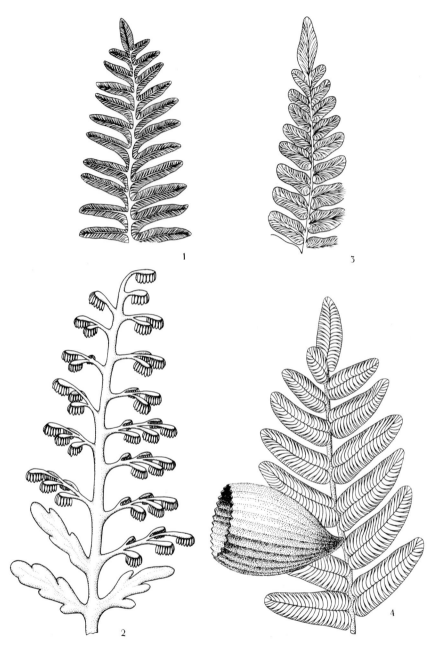

Figure 15. Seed ferns of the Pennsylvanian. 1. *Alethopteris;* 3. *Neuropteris,* approximately to scale (from Darrah); 2. *Crossotheca* sp., restoration of Illinois specimen (X2) (after Andrews); 4. Neuropterid type with reproductive body. Specimen from Holland (X2) (after Andrews).

Figure 16. The living ginkgo, *Ginkgo biloba*. Leaves and seeds are similar to fossil types which became extinct in North America and Europe during the Middle and Late Cenozoic. The ginkgo has been reintroduced by man in these areas (X1/2). (After Andrews.)

Figure 17. A Cretaceous angiosperm leaf, *Aspidophyllum* (approximately X 1/2). (After Darrah.)

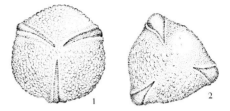

Figure 18. Pollen grains from angiosperms. 1. beech; and 2. oak. Originals approximately .03mm. in diameter. (After Wodehouse.)

microcells (Fig. 18). These remarkable reproductive parts are readily preserved in nonoxidizing environments and fossil pollen as well as spores from more primitive types are exceedingly abundant in rocks from Paleozoic to Recent. Their study is so specialized that a specific area of study called "palynology" has developed. Certain types are useful in biostratigraphic work, and their study for paleoclimatological determinations is especially important (Maher, 1972).

PLANTS AND CLIMATES OF THE PAST

Plant environment has special importance for the interpretation of Earth History. Most higher plants are rather sensitive indicators of their environments. Characteristic floras are found only in certain ecologic systems. Today's floras can be interpreted in terms of their present ecologic requirements and these same requirements can be assumed for like floras of previous time intervals. Thus, the breadfruit, *Artocarpus*, today lives only within 20 degrees latitude from the Equator. Since this plant is found fossilized in the Upper Cretaceous strata of Greenland which is above 60 degrees north latitude, obviously climatic conditions in Greenland (or the latitude of Greenland) have undergone drastic changes since that period. Eocene floras of California and Oregon are most similar to those found today in Central America where 80 inches of rainfall a year is common, and where there are no freezing temperatures. Again, climatic changes are evident. In this way, floras may be utilized for determinations of climates of the past.

REFERENCES

Andrews, H. N., Jr. 1961. *Studies in paleobotany.* New York: John Wiley and Sons.

Darrah, W. C. 1960. *Principles of paleobotany.* 2d ed. New York: Ronald Press.

Maher, L. J., Jr. 1972. Absolute pollen diagram of Redrock Lake, Boulder County, Colorado. *Quaternary Research* 2:531–53.

Wodehouse, R. P. 1959. *Pollen grains.* New York: McGraw-Hill.

Chapter Four

Fossil Invertebrates

Animals without backbones (invertebrates) evolved during the late Precambrian but were not abundant until approximately 600 million years ago. From that time until now, millions of species of invertebrates have lived and died, leaving a fantastic record of evolution.

There are approximately a dozen phyla of geologically important invertebrates. Some 80 percent of fossil animals are invertebrates. More students study invertebrates and more is known about this category than is known about any other fossil group. They probably are the most useful group in the interpretation of Earth History. A brief description of the major categories follows. Their stratigraphic range is shown in Figure 19.

PHYLUM PROTOZOA

This group of one-celled animals includes many structurally simple and "primitive" animals, but certain members of the group are neither simple nor primitive. Most protozoans have no hard parts and one would not anticipate finding a fossil *Amoeba* or *Euglena*. There are several groups which possess a shell structure and one of these, the Foraminifera, is geologically important.

Foraminifera

Most Foraminifera (Fig. 20.1–9) have a shell less than a millimeter in size which is an original secretion of calcium carbonate ($CaCO_3$) or may consist of organic material, or may be formed by the agglutinization of debris and held together by one of several cementing agents. Free floating and swimming, as well as bottom dwelling types of Foraminifera are known, and include both fresh water and marine types, although most are the latter.

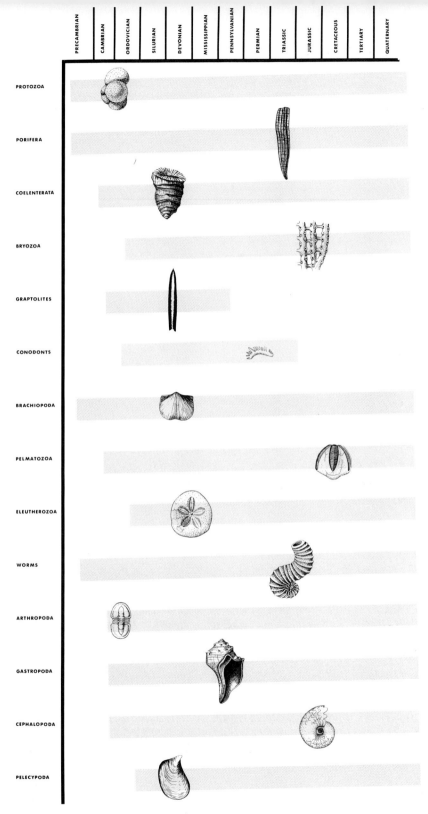

Figure 19. Development of the invertebrates and their distribution through time.

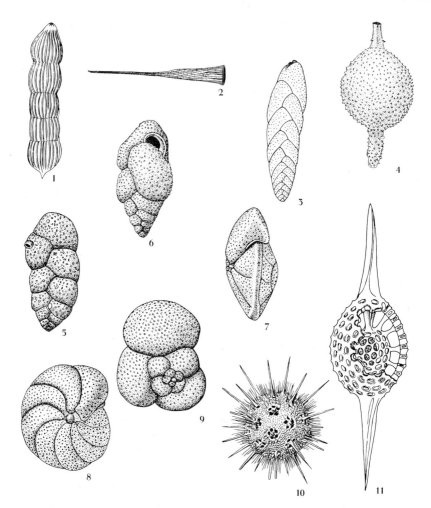

Figure 20. Fossil protozoans; Foraminifera, radiolarians, and tintinnids. 1, 3–9 are Foraminifera; 1. *Dentalina crosswickensis,* a straight five chambered form (X46), Early Cenozoic; 3. *Loxostoma plaitum,* chambers arranged side by side as well as vertically (X61), Cretaceous; 4. *Dentalina pseudoaculeata,* straight bulbous type (X98), Early Cenozoic; 5. *Gaudryina monmouthensis*—notice location of aperture below shell apex (X53), Cretaceous; 6. *Chiloguembelina crinita* —notice large aperture laterally situated (X141), Early Cenozoic; 7, 8. side and front views of *Cibicides harperi,* planispirally coiled form (X 80), Cretaceous; 9. *Globigerina inaequispira,* spiral arrangement of chambers (X98), Early Cenozoic (adapted from Olsson); 2. a Recent tintinnid, *Rhabdonella conica* (X100); 10, 11. radiolarians. 10. globular type with spines, *Haeckeliana darwiniana,* Recent (X50); 11. elongate type with outer shell broken to expose inner capsule, *Stylatractus giganteus,* Recent (X100) (after Campbell and Moore).

Other groups of protozoans have some geologic importance including the radiolarians (Fig. 20.10, 11) and tintinnids (Fig. 20.2). These protozoans have a silica (SiO_2) shell and average less than one millimeter in size. Most radiolarians were free-floating organisms and their shells have been useful in correlating Cenozoic sediment cored from the present deep seas. Several thousand species have been described.

PHYLUM PORIFERA

The sponges are the simplest multicellular animals. They are sessile, aquatic, variable in shape and are commonly colonial. They evolved from collared protozoans. Most living and fossil sponges are small (only a few inches or less in diameter), but forms with a body diameter of more than three feet are known. There probably are 2,000 living species, and approximately 2,000 fossil species are known. Because of their porous bodies, sponges have not been preserved as readily as other invertebrate groups and except in locally abundant areas, they are considered to be rare fossils.

Sponges do not have a rigid external skeleton but a rather soft three-layered body, the inner layer of which may contain a skeleton composed of siliceous, calcareous, or organic spicules (Fig. 21.1). The body generally is rounded and porous and water enters the hollow center through the pores and exits through a central opening often called the "mouth." In the process food is picked up by the cells of the inner wall. The skeleton, when present, may be firmly formed of interlocking spicules (Fig. 21) or the spicules may be loose in the flesh and serve as supports. Scattered spicules are more common than a whole skeleton.

PHYLUM COELENTERATA

This large and diverse group includes the corals, the jellyfishes, the hydras, and their kin. Some are microscopic and others range to 6 feet in length and may have tentacles 130 feet long. Even with this diversity in size it is curious to note that most coelenterates are constructed on a uniform pattern. This pattern is a simple sack-shaped body with a three-layered wall surrounding a large gastrovascular cavity (stomach-like structure) which contains no internal viscera. The mouth is located at one end and is surrounded by a group of tentacles, usually in cycles of two, four, six, or eight. Several important coelenterate types have skeletons (Figs. 22, 23) while others have no hard parts at all.

The largest and geologically most important group of coelenterates are the corals and their kin. More than 6,000 living species are known, all marine. This group includes the corals, sea pens, and sea anemones. Geologically, the most significant corals (= anthozoans) are the rugose and tabulates of the Paleozoic. Most are small and cone-shaped, and the

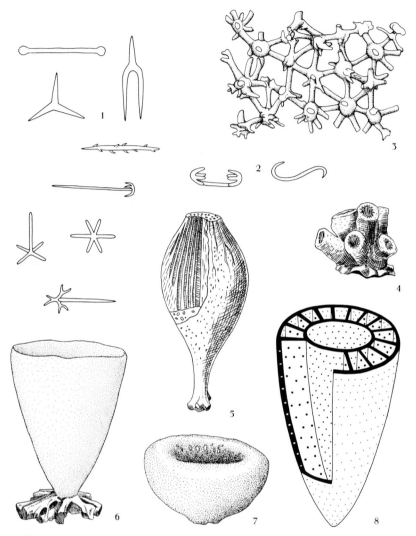

Figure 21. Sponges, sponge spicules, and sponge-like animals. 1. eight kinds of spicules of siliceous sponges, the large variety (×100); 2. two kinds of small siliceous sponge spicules (×1,000); 3. united spicule group of the sponge *Astylospongia praemorsa* from the Silurian of Germany (×100); 4. a calcareous sponge, *Eusiphonella bronni* from the Jurassic of Europe (×1); 5. wall structure in *Acrochordonia ramosa* from the Cretaceous of Germany (×30); 6. vase-shaped siliceous sponge from the Cretaceous of Europe (×1/2); 7. bowl-type sponge from the Silurian of Germany (×3/4); 8. diagrammatic view of an archaeocyathid sponge-like animal with inner and outer walls and vertical partitions. Note inner cavity and numerous pores (sketch greatly enlarged). (From Okulitch and de Laubenfels.)

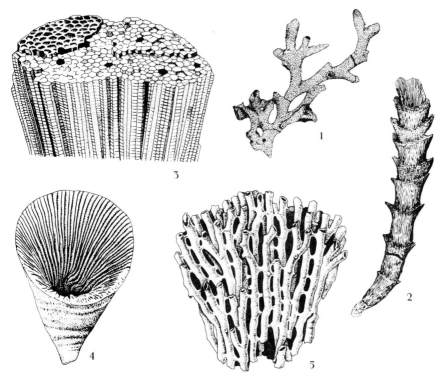

Figure 22. Corals of the Paleozoic. 1. *Alveolites winchellana,* surface view of a tabulate coral from the Devonian of Kentucky (X1/2); 2. *Blothrophyllum greeni,* a rugose form showing evidence of pauses in growth, from the Devonian of Kentucky (X1/2); 3. a large tabulate colony showing cross section of individuals and transverse tabula; *Favosites gothlandicus,* Silurian (X1); 4. *Kionelasma conspicuum.* A rugose coral showing vertical septa in cup, Devonian of Kentucky (X1); 5. a tabulate colony showing discrete yet connected individuals, *Syringopora ramulosa,* Lower Carboniferous of Belgium (X1). (1, 4, 6, after Bayer et al., 2, 3, 5, after Stumm.)

most important skeletal features are the radial plates, called septa, which project from the wall of the "cone" to the center of the specimen (Fig. 22.2–4). In many types, additional plates called tabulae grow transversely to the septa (Fig. 22.3).

Fossil corals have had two rather separate and distinct histories: the first, a Paleozoic history, characterized by species that largely were extinct by the end of the Paleozoic, and second, a post-Paleozoic history, including the evolution of modern types.

The types illustrated in Figure 22 became widespread in the Silurian, declined until another evolutionary burst occurred in the

Devonian and then there was another decline in number until they became extinct in the Late Permian.

After the Paleozoic the scleractinians, or stony corals (Fig. 23), became abundant. They first appeared in the Middle Triassic. Whether they evolved from a rugose ancestor or from a separate line of Paleozoic anemones without shells is unknown. All of the modern "true" corals, so common on the flourishing reefs in the temperate marine waters of the world, are of this type. Today, this group grows best in water at

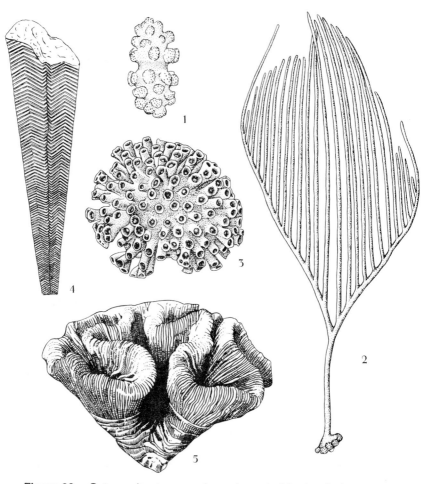

Figure 23. Octocorals, stony corals, and conulariids. 1. spicule, approximately .05 mm., from a modern octocoral; 2. *Ctenocella pectinata* (approximately X1/4), Pacific form; 3. a modern stony coral colony, *Duncanopsammia axifuga,* (X1/4) from Australia; 4. *Paraconularia worthi,* a Permian conularid from India (X1); 5. modern stony coral colony, *Trachyphyllia geoffroyi* (X1/2), Australia. (After Bayer et al.)

approximately 25°C. Water is needed which is shallow enough to allow plenty of sunshine to penetrate (less than 600 feet) and in which a moderate amount of water circulation occurs. Coral reefs, which are mound-like structures built by organic activity, are an impressive part of the geologic record. During the Silurian, massive as well as smaller reefs grew in the seas which covered much of the area of Iowa, Wisconsin, Illinois, and Indiana. Also, Devonian and younger reefs are found in many parts of the world.

PHYLUM ECTOPROCTA (BRYOZOA)

One of the least conspicuous major groups of invertebrates is the group of sessile (attached) animals known as bryozoans or ectoprocts. These tiny colonial animals are commonly encountered as encrustations on rocks or on other shells and certain types have the appearance of a perforated plate or sieve (Fig. 24). Each tiny opening of the "sieve"

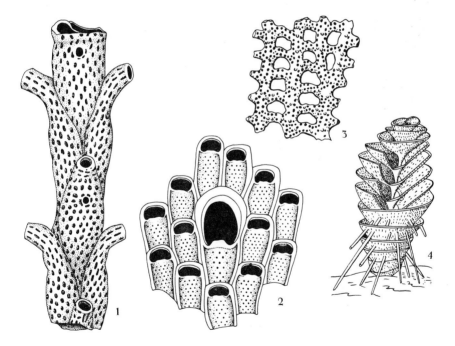

Figure 24. Ectoprocta (Bryozoans). 1. *Tubucellaria vicksburgica,* mode of growth in Eocene colony from Gulf Coast (X25); 2. *Steginoporella jacksonica,* individuals of colony with large apertures, Eocene, Gulf Coast (X25) (after Cheetham); 3. *Septopora subquadrans,* small openings of individuals and larger "windows" through colony (X5), Mississippian, North America; 4. *Archimedes* sp., a diagrammatic illustration of lace-like frond around a stony axis, Mississippian-Pennsylvanian of North America (X1) (after Easton).

houses a single individual whose soft parts are lost after death. The houses are little more than a cave in a calcareous mass but they are all that remain for paleontologic study.

There are approximately 3,000 living species and these animals were equally abundant during the geologic past.

GRAPTOLITES

The degree of understanding that we have for most fossil groups is based largely on our understanding of living representatives of the particular group. This dependency between fossil and living types is well illustrated when we deal with extinct groups such as the graptolites (Fig. 25). No form of life living today has the same structure as the

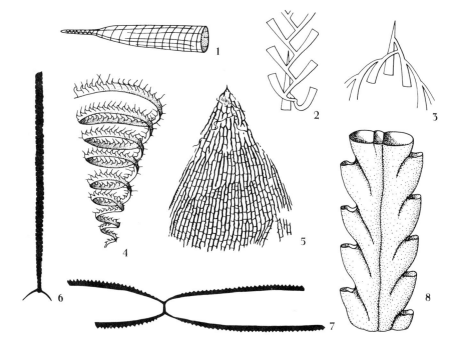

Figure 25. Graptolites. 1. embryonic part of graptolite colony from which budding of individuals occurs, the prosicula (X50); 2 and 3. two growth patterns from prosicula, one showing downward or primitive pattern, the other upward or advanced; diagrammatic; 4. an unusual graptolite colony, *Monograptus turriculatus* from the Lower Silurian of Bohemia (X2); 5. a primitive colony showing numerous branches and downward direction of growth, *Dictyonema flabelliforme* (X1/2); 6. an advanced type, *Climacograptus bicornis* (X1), Middle Ordovician, illustrates most typical preservation pattern; 7. colony with four branches, *Tetragraptus approximatus* (X2), Ordovician of Sweden; 8. enlarged part of *Glyptograptus* colony to illustrate individual cups of animals and growth relationship (X15). (After Bulman.)

graptolites, and although a group of pterobranches, vertebrate-related "chordate-like" animals, are similar, some authorities insist that the similarities are only superficial and that the graptolites, which became extinct in the Paleozoic, are unique.

We know that graptolites were colonial and marine. We know that they produced an organic (chitinoid) external skeleton which consisted of a series of cups or chambers arranged along a central stalk (Fig. 25.8) and that each cup housed an animal. Most fossil specimens are little more than carbon impressions, although distinctive shapes are known (Fig. 25.5, 6, 7).

Both sessile and floating types are known but most of the geologically important types were floating colonies. This also seems to explain the widespread abundance of graptolites in black shale, sediment deposited under conditions apparently inhospitable to most forms of life.

CONODONTS

Another group of fossils whose precise biologic relationship remains a mystery is the conodonts. Conodonts were animals whose bodies bore microscopic-sized elements which are cone-like, and comb-like but generally unlike anything living today (Fig. 26). These elements are abundant and widespread; different kinds seem to characterize each different age rock so that their value as index fossils is not excelled by any other group.

Conodont elements are composed of calcium phosphate and, though microscopic, are readily recovered from acid residues of limestones and shales. Certain types occur together as sinistral (left-hand) and dextral (right-hand) pairs. Other types are bilaterally symmetrical. In addition, certain assemblages showing patterns of pairs have been discovered suggesting that the individual conodont element was part of some bilaterally symmetrical soft-bodied animal. Impressions of fossils thought by some to be the whole animal have been found in Upper Paleozoic rocks in Montana recently. Their interpretation is a matter of considerable interest.

PHYLUM BRACHIOPODA

The brachiopods constitute one of the most important groups of fossil invertebrates. The soft parts of these "clam-like" animals are enclosed by a shell which consists of two valves which are dissimilar in shape (Fig. 27). The two valves are hinged at one end, which is referred to as the posterior. Most brachiopods live semi-attached to some substratum by a large muscle which emerges from an opening at the posterior end. The valves at the anterior part of the shell can be opened for feeding and respiratory functions.

Figure 26. Ontogenetic (individual growth development) and evolutionary development of conodont, *Ancyrodella*. Earliest growth stages are shown in *a* to *g*. Three different lines of evolution (adult forms) are shown in *g–m*, *g–s*, and *g–w*. Notice different arrangement of surface nodes on these internal skeletal elements of an Upper Devonian conodont (X30) (after Müller and Clark).

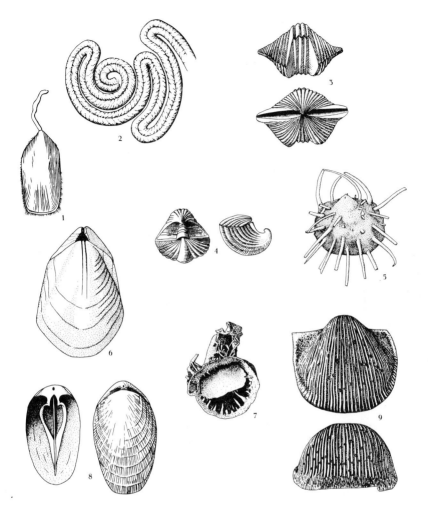

Figure 27. Brachiopods and trace fossils. 1. living type from Hawaiian Islands, *Lingula hawaiiensis* (X1); 2. a trace fossil (possible worm trail), *Taphrhelminthopsis auricularis,* Cenozoic of Italy (X.2); 3. tail end and beak views of articulate *Platystrophia acutilirata* (X1), a common fossil in Ordovician rocks of Midwestern United States; 4. beak and side views of *Cyrtina alpenensis,* articulate from the Devonian of Michigan (X1); 5. *Echinauris lateralis,* a Permian articulate with clasping spines (X4); 6. internal mold of *Pentamerus laevis,* a Silurian articulate from North America (X3); 7. an unusual articulate with coral-like shell, *Prorichthofenia permiana* from Permian of Texas (X1); 8. interior of one valve and exterior view of articulate *Rensselaeria marylandica,* Devonian of Maryland (X1); 9. top and end views of *Labriproductus worthensis,* a Mississippian articulate from Missouri (X2). (1, after Moore, Lalicker, and Fischer; 2, adapted from Hass et al.; 3, 4, 6, 8, after Easton; 5, 7, 9, adapted from Muir-Wood and Cooper.)

More than 15,000 species of fossil brachiopods have been described, and approximately 200 species are living today. Brachiopods generally are small, ½ to 3 inches in their greatest dimension. The two-valved shell is either calcareous or a mixture of calcareous and organic material and is readily preserved. The brachiopod consists of a soft body covered with a fleshy membrane which also secretes the shell. Food is taken by a lophophore, or arm-like structure, located in the anterior part of the shell.

The shell form of brachiopods is diverse (Fig. 27). Robust, flat, triangular, elongate, wide, and narrow shapes are known, and shells may be smooth or ornamented with coarse ribs and spines.

WORMS

Even worms, the majority of which have no hard parts, have left a record in the rocks of the earth. Worm tubes, body impressions, and quite often the jaws (Fig. 28) of certain types are found in rocks ranging in age from the late Precambrian to the Recent. The worm jaws are commonly well preserved, but the body impressions and tubes are much less so and are difficult to find and to distinguish. Living types collectively called worms are actually so diverse in their structure that several distinct phyla are recognized by specialists. These separate classifications are not of great value to the paleontologist who must deal only with preservable parts or traces of those parts.

Figure 28. Part of a jaw (scolecodont), from *Idraites,* a worm from the Ordovician of Wisconsin. Scolecodonts are usually the only part of worms fossilized (X14).

PHYLUM MOLLUSCA

Snails, clams, and cephalopods are three principal members of this important group. Most molluscs have a calcareous shell and those which lack this external armor are considered specialized. In addition to the three major groups, there are several other groups whose biologic interest is far greater than their geologic importance. This category includes the chitons, the monoplacophorans and scaphopods, and a few other extinct groups.

Most molluscs live in a marine environment and fossil types confirm that this has been the pattern of the past. There are fresh water types, and terrestrial air-breathing snails are not uncommon, however.

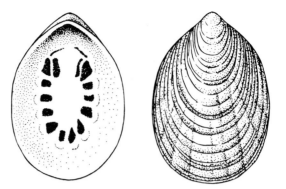

Figure 29. *Pilina,* a Silurian monoplacophoran, internal and external view of calcareous valves. This is a representative of a group of molluscs that show characteristics of symmetry linking it with annelid worms as well as with advanced molluscs. Silurian (X.7). (After Knight et al.)

The range of environments is rounded out by certain cephalopods (squids) known today which can "fly" 150 feet through the air.

The earliest molluscs appear in the Cambrian and the group has continued as one of the most important groups of invertebrates to the present time. More than 125,000 species of molluscs have been described, of which approximately 80,000 are living.

The primitive molluscs are called monoplacophorans (Fig. 29). Once thought to have been extinct for 400 million years, living species have now been found in both Atlantic and Pacific oceans. Monoplacophorans are bilaterally symmetrical and probably were ancestral to the other molluscs. They apparently evolved from the annelid worms.

Class Gastropoda

Gastropods, or snails as they are commonly called, are univalved molluscs whose shell may be cap-like, coiled, or tubular (Fig. 30). The spiral coiling type is most common but one extinct group was coiled in one plane (planispiral), and there are fossil and living types with only traces of coiling preserved (the patellids, limpets, abalones, etc.) Living gastropods have asymmetrical body structures, and observation of the ontogeny (individual growth development) of many species has convinced specialists that an evolutionary developmental pattern of twisting from an original symmetrical condition took place.

The shell consists of a spire (all volutions above the body chamber), and a body chamber (the last complete volution). The shells are commonly ornamented with ridges, nodes, spines, etc., and they may be low or high coiled or show no coiling at all.

Gastropods are the only mollusc group to have attained a terrestrial habitat. The majority of snails, however, are marine or fresh water animals and live at depths of less than 200 feet. Most are active swimmers, crawlers, or drifters and only a few are sessile. They are either

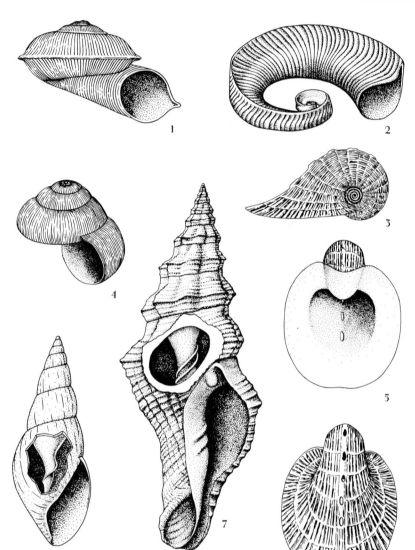

Figure 30. Gastropods. 1. low-coiling type, *Pleuromphalus seductor,* Silurian of Bohemia (×1.3); 2. loosely coiled type, *Ecculiomphalus alatus,* oblique view, Ordovician of Europe (×3/4); 3, 5, 8. planispiral variety, *Tremanotus alpheus,* side, inner, and upper views of a gastropod showing change in growth at maturity, Silurian, New York (×.5); 4. *Straparollus* sp., common Mississippian type (×3/4); 6. *Soleniscus typicus,* a high spired type with opening to show central axis, Pennsylvanian, Illinois (×1); 7. a high spired gastropod with opening to show internal axis, *Latirus lynchis,* Miocene, France, approximately (×1). (After Knight et al.)

herbivores or carnivores and certain types are particularly fond of oysters.

Class Cephalopoda

The squids, octopuses, cuttlefish, argonauts, and pearly *Nautilus* are the living representatives of a molluscan class which also includes a host of extinct forms. Members of this group have lived in abundance since the Late Cambrian. The 650 living species are considerably different from the fossil ancestors whose greater numbers suggest that cephalopods found life more comfortable in the geologic past. Cephalopods have always been marine organisms and most representatives of the group have been agile, active swimmers with a carnivorous diet. Living cephalopods are so active and shy that their study has been very difficult and until recently little has been known about their habits and habitats. The *Nautilus* is the only living cephalopod which carries its chambered shell externally (Fig. 31). Most other living types are naked, although a

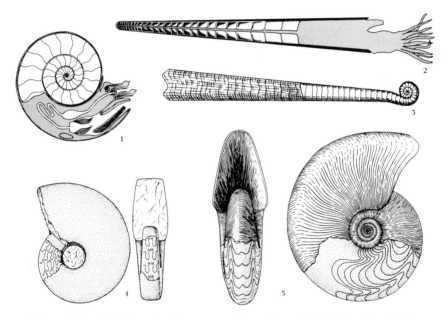

Figure 31. Cephalopods. 1. cross section of hypothetical fossil cephalopod with body parts, based on analogy with living *Nautilus;* 2. reconstruction of section of straight nautiloid showing chambers and soft parts, 3. an Ordovician straight nautiloid from the Baltic, *Lituites litmus* (X.3); 4. *Prouddenites primus,* lateral and end view of a Pennsylvanian ammonoid; shell removed to show suture (X1.5); 5. typical Devonian ammonoid, *Manticoceras* sp., with shell and suture, an end and lateral view, approximately (X1). (1, 4, 5, after Arkel et al.; 2, 3, after Moore, Lalicker and Fischer.)

few have a small internal shell. Most fossil types were similar to the *Nautilus* in shell-bearing habit.

The shell of both living and fossil types of the *Nautilus* and its relatives is univalved and internally partitioned.

The internal siphuncle (an internal tube that runs from the first formed chamber to the body chamber) and the configuration of the suture line (contact of chamber wall and outer wall) are important in the classification of fossil cephalopods. One group, including the modern *Nautilus* which possesses relatively simple suture configurations and highly variable siphuncles, is called the nautiloids (Fig. 31). Another group, which ranged from the Middle Paleozoic to the end of the Mesozoic and had elaborate suture lines and a fairly uniform siphuncle pattern, is called the ammonoids (Fig. 32).

The squids, octopoids, and other cephalopod types, largely without shells, have left a very poor geologic record but are known in Pennsylvanian rocks in Illinois and in Mesozoic and Cenozoic rocks in several other parts of the world.

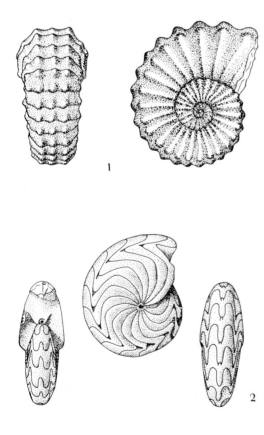

1

2

Figure 32. Ammonoid cephalopods. 1. advanced type with ribs and spines, suture not shown, *Acanthoceras rhotomagense,* Cretaceous, France (X1/3); 2. end and lateral views of *Imitoceras rotatorium,* outer shell removed to show suture (X1), Mississippian, Indiana. (1, 2, after Arkel et al.)

Class Pelecypoda (Bivalvia)

The pelecypods, which include clams, oysters, and scallops, have a shell consisting of two valves (Fig. 33). The valves are hinged on the dorsal side of the animal and hence a left and right side are produced. Shell structure in this group is diverse and seems to reflect the animal's living habits to a greater degree than it does its kinship to other kinds. Hence, burrowers commonly are elongate and smooth, sessile forms normally are scabious, and free swimming types are more symmetrical. It is difficult to relate such diverse types in understandable lineages, and most shells give only a hint of the former soft-bodied animal which inhabited them. A majority of fossil pelecypods, as well as those living today, were shallow marine types, but fresh water varieties are also abundant.

One of the most important groups is the oysters and their kin. These were (and are) sedentary attached forms in which considerable

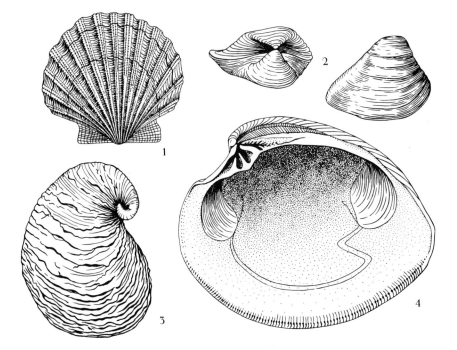

Figure 33. Pelecypods. 1. *Lyropecten madisonius,* a Miocene type from North America (X1/3); 2. two views of *Corbula undifera,* Cretaceous, Wyoming (X1); 3. a Cretaceous oyster-like shell, *Exogyra ponderosa* (X1/2), Texas; 4. internal view of a valve of *Mercenaria mercenaria* (X1), Cenozoic and Recent, world wide. Notice muscle scars at either end of shell. (Adapted from Easton.)

modification of the shell has taken place. The left valve is the lower attached one and attachment is normally by cementation. The right valve may become a small, lid-like structure. The oysters had their origin during the Permian and descendants were common during the Mesozoic and Cenozoic, as well as being abundant and of great economic importance today.

PHYLUM ECHINODERMATA

The animals which are classified together as echinoderms are outwardly different and one might suspect that there is little similarity between the starfish and the stalked "sea lily," or crinoid. Yet all members of this diverse group possess marked similarity in basic structure. All have a primitive bilateral symmetry, which is often well developed during early embryonic growth but which, by maturity, is masked by a radial or "five-fold" symmetry. All have some kind of a water-vascular system which may combine functions of food gathering and locomotion with respiration. This system, with specialized internal structures and external plates, is the most distinctive feature of the phylum. Most, but not all, have an external armor of calcite plates.

Echinoids
The sea urchin family includes hundreds of living types and an even greater number of fossil species. Many have their mouth on a flattened lower surface and their anus on the opposite or upper side (Fig. 34.1, 2, 4). Living types have a shell with five double rows of plates which run from pole to pole. This twenty-plate arrangement is standard among living forms but the oldest fossil types are noteworthy because the number of plates was not standardized.

Stelleroids
The starfishes and their kin have been living in the oceans for more than 400 million years. There are two principal types: the asteroids, or true starfishes, whose body and arms are not clearly separated (Fig. 34.5); and the ophiuroids, including the brittle stars, serpent stars, and basket stars, whose slender arms are more obviously differentiated from the body (Fig. 34.3). Fossil types are very similar to those living today. The individual elements of the starfish body are readily scattered after the death of the animal and the disintegration of the muscles. Because of this, a whole fossil starfish is rare.

Holothurians
These sea cucumbers have elongate "cucumber" shaped bodies. They are different from other echinoderms because they possess no external skeleton of plates but instead may have tiny calcite sclerites, or spicules, embedded in their flesh (Fig. 35).

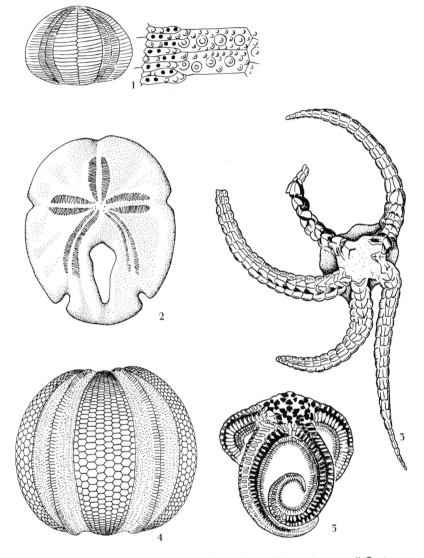

Figure 34. Echinoderms; echinoids, and starfishes. 1. a small Cretaceous echinoid with enlargement of plates showing spine bases and pore openings (X.7); 2. a Pliocene sanddollar (echinoid), *Encope macrophora* (X1/2), South Carolina; 3. *Hallaster cylindricus,* an Ordovician ophiuroid from Canada (X.3); 4. Mississippian echinoid with numerous rows of plates, *Melonechinus multiporus,* Missouri (X1); 5. a Pennsylvanian starfish, *Calliasterella mira,* Russia (X1). (1, 2, 4–6, after Moore, Lalicker and Fischer; 3, after Easton.)

Figure 35. Echinoderm. Holothurian skeletal parts (sclerites), hook, anchor, and wheel types. These are Mesozoic and Cenozoic varieties (X 180). (After Moore, Lalicker, and Fischer.)

Crinoids

Living echinoderms include approximately 800 species of crinoids, or sea lilies, whose greatest concentration apparently is in the southwest Pacific-Indian Ocean area. A far greater number and variety of sea lily species lived in the past, and recent evidence suggests that in the evolution of the group, many forms attained a free-moving mode of life. The crinoid body consists of two main parts: a calyx, or globular shaped "head," which contains the visceral organs, and which has arms which extend upward for food gathering; and a stem, or series of plates, which serve to hold the "head" or calyx above the sea floor (Fig. 36.5, 6, 8). There may be an attachment structure. The presence of a stem and the brilliant colors which may show in the membraneous covering of the calcareous plates has led to the present "sea lily" designation. Many living types are free-moving and some use their arms for rapid movement along the bottom.

Primitive Forms

Echinoderms were more diversified and numerous during the Early and Middle Paleozoic than they are today. Extinct types include a variety of forms (Fig. 36.1, 3, 4, 7). One primitive type (Fig. 36.2) had a symmetry which approached a bilateral condition, a tail-like appendage, an oral-aboral elongation, and an arrangement of plates which has led some students to consider these as the possible ancestor of the vertebrates. Should this or another echinoderm-vertebrate transition be established, a major link in the chain of life will be understood.

PHYLUM ARTHROPODA

This phylum includes the crabs, lobsters, spiders, insects, and other living invertebrates, as well as the extinct trilobites and eurypterids, all of which have segmented bodies and jointed limbs. The typical arthropod body is elongate and bilaterally symmetrical with vital organs enclosed in an organic or chitinoid skeleton. In many arthropods such as the shrimp, the skeleton is flexible, but in others, such as the lobsters and crabs, it is hard and brittle because of the presence of calcium

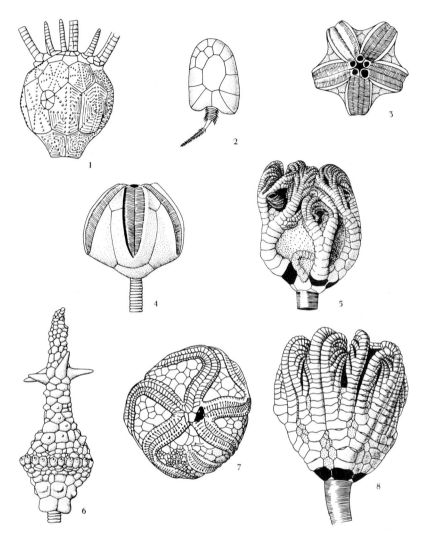

Figure 36. Echinoderms. Crinoids, blastoids, carpoids, and cystoids. 1. the cystoid *Hemicosmites pyriformis* from the Ordovician of Russia (approximately X1); note pore pattern on plates; 2. *Mitrocystites mitra,* a carpoid from the Ordovician of Bohemia, mouth at opposite end from "stem-like" appendage (X1); 3, 4. *Pentremites sulcatus,* a Mississippian blastoid in top view (3) and lateral view (4). Openings at top related to internal water system (X1); 5. Mississippian crinoid, *Onychocrinus ulrichi,* with arms folded over calyx, (X1); 6. crinoid with arms removed to show unusual calyx, *Uperocrinus nashvillae,* from Mississippian of Tennessee (X1); 7. primitive echinoderm, *Edrioaster bigsbyi,* Ordovician of Canada (X1); 8. *Talanterocrinus* sp., a Pennsylvanian crinoid with arms rising from calyx (X1). (3, 4, adapted from Easton; others from Moore, Lalicker and Fischer.)

carbonate. Those skeletons which have been impregnated with calcium carbonate have left the best fossils.

The group is enormously large. More than 700,000 species of insects are known. When this number is added to approximately 40,000 species of spiders and their kin, crustaceans, and millepedes, the total amounts to almost 80 percent of the 1 million living and fossil species of vertebrates and invertebrates. Geologically, the group is not as important as the present living numbers might suggest, and except for the trilobites of the Paleozoic, small ostracodes, and a few crustaceans, fossil arthropods are not common.

Arthropods have been highly successful in diverse habitats which range from the deep seas to high mountain latitudes. Some are aerial, some aquatic, and a great many are terrestrial.

Trilobites

The trilobites are one of the two most important groups of geologically significant arthropods. The group is characterized by the possession of a skeleton (carapace) which is divided into three lobes. The head, the thorax, and the tail comprise the transverse segmentation but these structures are also lobed longitudinally. On the underside of the trilobite, biramous or double appendages existed in life, a pair for each segment of the thorax (Fig. 37). The visceral organs are poorly known and inferences as to muscular, nervous, or respiratory systems or reproduction are based on information gained from study of living relatives. Trilobites were numerous during the Early Paleozoic but were extinct by the end of that Era.

Ostracodes

The second most important group of geologically significant arthropods is the ostracodes (Fig. 38.3). In contrast to the trilobites, ostracodes are abundant today and the study of living types has enhanced our knowledge of the fossils.

Ostracodes are small animals, most of which have a two-valved shell. The body is rarely preserved but numerous shells, only a millimeter or so in size, have been found in rocks ranging from the Cambrian to the Recent. They are abundant in both marine and fresh water areas today and one terrestrial species is known. Living ostracodes molt periodically (as do other arthropods) and up to nine molting stages have been noted.

Insects, Arachnids, and Eurypterids

Of less direct geologic importance but of considerable biologic significance are the arachnids, including the spiders (Fig. 38.1); the insects (Fig. 38.2); and a variety of living and extinct arthropods includ-

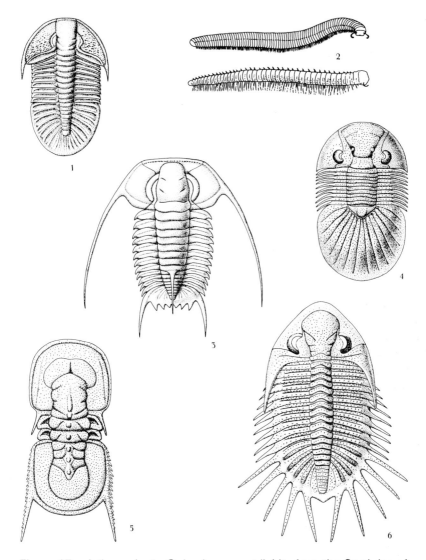

Figure 37. Arthropods. 1. *Orria elegans,* a trilobite from the Cambrian of Utah, well defined head but thorax and tail less distinctly separated (X.5); 2. upper, *Lulus* sp. a living millepede compared with *Euphoberia* sp. a fossil millepede from the Pennsylvanian of Illinois (both approximately X1); 3. *Pseudokainella keideli,* an Ordovician trilobite with long spines and median spine in central lobe (X 6); 4. *Scutellum costatum,* a Devonian trilobite illustrating eye bases, small thorax and large tail (X.5), Germany; 5. *Pleuroctenium granulatum,* a Cambrian trilobite without eyes showing more or less equal sized tail and head and small thorax (X4.5); 6. *Asteropyge punctata,* with large spinose tail, Devonian of Germany (X1.5). (2, from Moore, Lalicker and Fischer; others from Harrington et al.)

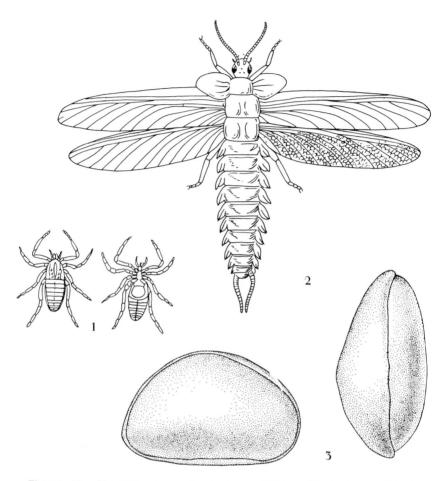

Figure 38. Miscellaneous arthropods. 1. Spider, *Plesiosiro madeleyi*, Pennsylvanian, Europe (X1); 2. Winged insect, *Stenodictya lobata*, Pennsylvanian, France (X1); 3. ostracodes *Ovovytheridea nuda*, lateral and end views, Cretaceous, Africa (X60). (1, 2, from Moore, Lalicker and Fischer; 3, adapted from Benson et al.)

ing the eurypterids (Fig. 39). The eurypterids, or sea scorpions, were both small and large-sized arthropods which inhabited the brackish and marine waters of the Paleozoic. Their restricted environmental range has made them of only secondary importance in geologic studies.

Insects, spiders, etc., have left a poor geologic record. One remarkable insect with a wing span of 29 inches is known from the Pennsylvanian (Fig. 38.2). Insects must be considered of great geologic importance for their work in fertilization and spreading of pollen for plants.

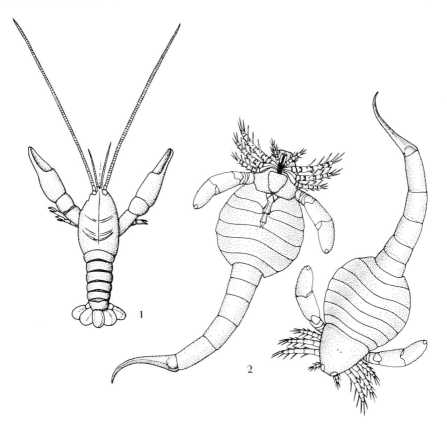

Figure 39. Eurypterids and lobsters. 1. *Eryma leptodactylina,* a lobster from the Jurassic of Germany (X1); 2. dorsal and ventral views of the eurypterid, *Carcinosoma scorpionis,* from the Silurian of New York. These sea scorpions included some of the largest invertebrates which have lived (X.13). (1, after Moore, Lalicker and Fischer; 2, after Stormer et al.)

REFERENCES

Arkell et al. 1957. Part L, Mollusca 4, Cephalopoda, Ammonoidea. In *Treatise on invertebrate paleontology,* ed. R. C. Moore. Geological Society of America and University of Kansas Press, Lawrence.

Bayer et al. 1956. Part F, Coelenterata. In *Treatise on invertebrate paleontology,* ed. R. C. Moore. Geological Society of America and University of Kansas Press, Lawrence.

Beerbower, J. R. 1968. *Search for the past—an introduction to paleontology.* Englewood Cliffs, N.J.: Prentice-Hall.

Benson et al. 1961. Part Q, Arthropoda 3, Crustacea, Ostracoda. In *Treatise on invertebrate paleontology,* ed. R. C. Moore. Geological Society of America and University of Kansas Press, Lawrence.

Black, R. M. 1973. *The elements of paleontology.* Cambridge: Cambridge Press (paperback).

Bulman, O. M. B. 1955. Part V, Graptolithina. In *Treatise on invertebrate paleontology*, ed. R. C. Moore. Geological Society of America and University of Kansas Press, Lawrence.

Campbell, A. S., and Moore, R. C. 1954. Part D, Protista 3. In *Treatise on invertebrate paleontology*, ed. R. C. Moore. Geological Society of America and University of Kansas Press, Lawrence.

d'Orbigny, Alcide 1840. *Paléontologie francaise: Terrains crétacés.* Paris: A. Bertrand, V. Masson.

Easton, W. H. 1960. *Invertebrate paleontology.* New York: Harper and Brothers.

Funnell, B. M., and Riedel, W. R., eds. 1971. *The micropaleontology of oceans.* London: Cambridge University Press.

Harrington et al. 1959. Part O, Arthropoda 1. In *Treatise on invertebrate paleontology*, ed. R. C. Moore. Geological Society of America and University of Kansas Press, Lawrence.

Hass et al. 1962. Part W, Miscellanea. In *Treatise on invertebrate paleontology*, ed. R. C. Moore. Geological Society of America and University of Kansas Press, Lawrence.

Knight et al. 1960. Part I, Mollusca I. In *Treatise on invertebrate paleontology*, ed. R. C. Moore. Geological Society of America and University of Kansas Press, Lawrence.

McAlester, A. L. 1968. *The history of life.* Englewood Cliffs, N.J.: Prentice Hall (paperback).

Müller, K. J., and Clark, D. L. 1967. Early Late Devonian conodonts from the Squaw Bay Limestone. *Journal of Paleontology* 41:902–19.

Okulitch, V. J., and Laubenfels, M. W. de. 1955. Part E, Archaeocyatha and Porifera. In *Treatise on invertebrate paleontology*, ed. R. C. Moore. Geological Society of America and University of Kansas Press, Lawrence.

Olsson, R. K. 1960. Foraminifera of latest Cretaceous and earliest Tertiary age in the New Jersey Coastal Plain. *Journal of Paleontology* 34:1–58.

Raup, D. M., and Stanley, S. M. 1971. *Principles of paleontology.* San Francisco: W. H. Freeman

Stormer, L., Petrunkevitch, A., and Hedgpeth, J. W. 1955. Part P, Arthropoda 2. In *Treatise on invertebrate paleontology*, ed. R. C. Moore. Geological Society of America and University of Kansas Press, Lawrence.

Tasch, P. 1973. *Paleobiology of the invertebrates.* New York: John Wiley and Sons.

Chapter Five

Fossil Vertebrates

Vertebrate animals have been found only in Ordovician and younger rocks. While their origin is obscure, the evolution of these animals from fish to mammals is remarkably well understood. Sequences of vertebrate fossils are known from many continents and there remain few missing links in the evolution of the major categories of vertebrates. The development of the vertebrate, or backboned, animals is one of the most fascinating parts of the story of evolution. Evolution of the major vertebrate categories is summarized in Figure 63. For the most part, the record of vertebrates is not as complete as it is for certain of the invertebrates.

ORIGINS

The origin of vertebrates and their ancestry are problems still obscured by too few facts. The earliest remains interpreted as true vertebrates come from Ordovician rocks. It appears fairly certain that the ancestors of the vertebrates are to be found among the invertebrates. Just which of several different invertebrate phyla were the ancestors is still not clear, however, there is some evidence that points to an unusual group of echinoderms known as carpoids (Fig. 36.2) as possible vertebrate ancestors.

There are five large categories of vertebrates: fishes, amphibians, reptiles, birds, and mammals. The ecologic adaptation of members of these groups to a variety of environments is well documented and the discussion that follows will treat the vertebrates in terms of their major ecologic classification.

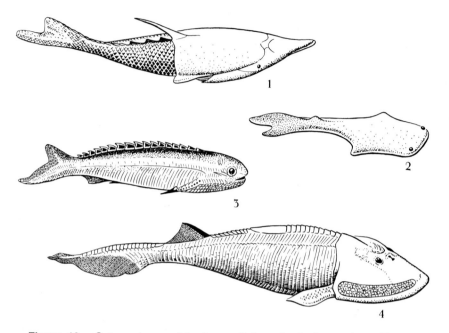

Figure 40. Ostracoderms of the Lower Paleozoic. 1. *Pteraspis;* 2. *Thelo-dus;* 3. *Pterolopis;* 4. *Hemicyclaspis.* Most ostracoderms had heavy head shields with scales behind, but variation from this "norm" is indicated. (All approximately X1/2.) (After Colbert.)

AQUATIC VERTEBRATES

Every major group of vertebrates has had aquatic members among its number. These include fishes, amphibians, swimming reptiles, swimming mammals, and swimming birds. The wide range of ecologic possibilities in the marine environment has successfully encouraged this development.

FISHES

There were at least four major groups of fishes living by the end of the Devonian and all except one are represented among modern fish faunas.

Agnatha

The agnathids or jawless fish are evidently the most primitive and were the first vertebrates to appear. Most of the Paleozoic types were heavily armored and are called ostracoderms. All of the armored types were extinct by the end of the Paleozoic and only the lamprey and its kin, which are considered by many to be primitive agnathids, have survived to the present.

Many ostracoderms were small and flattened. Their dorsally placed eyes and tail structure suggest that life was spent primarily as bottom-dwellers, rather than as active swimmers (Fig. 40). A few were lightly armored and perhaps more active.

Placoderms

The absence of jaws limited the ostracoderms to a rather narrow environmental range. Their nourishment depended upon mud-sucking, and they digested available organic particles from the muddy, aquatic bottoms.

The progressive development of jaws from the primitive gill arch apparatus is clearly suggested by the nerve arrangement, and by certain embryologic changes observed in living types. Fossil forms provide additional evidence of this development. The appearance of jaws was a major advancement for vertebrates and allowed them many new environmental niches and a greater variety of food (including each other).

Placoderms were a large and varied Paleozoic group, and like their probable agnathid ancestors, they inhabited both fresh and marine water environments (Fig. 41). Enormous marine placoderms were the largest vertebrates of their time. Placoderms were widespread, but

Figure 41. Placoderms and sharks. 1. *Bothriolepis,* a Devonian placoderm; and 2. *Cladoselache,* a Devonian shark. (Approximately X1/3.) (After Colbert.)

success, measured in the eons of geologic time, was short-lived and all became extinct during the Paleozoic.

Sharks

This large and successful group, which includes skates and rays, probably descended from the placoderms but has many structural features which distinguish it from the other fishes. Most distinctive of these differences is a skeleton of cartilage, not bone. Cartilage is not readily preserved; therefore a calcified brain case, rare body impressions, body denticles, or more commonly, teeth, are about the only part of sharks preserved for paleontologic study. General body shape has changed little since their appearance in the Devonian (Fig. 41.2) and a predaceous mode of life probably always has characterized the group.

Most living types are marine, but a fresh water group lived during the Paleozoic. The sharks appear to have been a rather uniform group from their first appearance to the present day, although reduction in number of gill slits, modification of jaw suspension, and narrowing of the fin base for greater mobility, have been evolutionary changes.

Bony Fishes (Osteichthyes)

The bony fishes, including most of the types familiar to the sportsman, are the "true" fishes and they have continued to increase in kinds and numbers from the Devonian to the Recent. The earliest forms were evidently fresh water types but their descendants inhabited the Paleozoic seas as well as lakes in tremendous numbers.

The earliest bony fishes were small forms with thick scales, large eyes, and fins, some of which had bony spines. This primitive type was almost extinct by the end of the Paleozoic although a few types survive today. Several important categories appeared during the Cenozoic and more than 30,000 species are known today.

The most important group from the point of view of evolution are the lobe finned fishes, the immediate ancestral group of all the other vertebrates.

The lobe finned fishes appeared in the Devonian. They had an air sac connected to the throat, a feature present in all bony fish, and also internal nostrils. It is assumed that the primitive air sac had an unstable position on the throat of the fish; unfavorable if it were to be utilized for air breathing. Through time, the position and stability of the air sac changed. In modern bony fish, the sac is used as a hydrostatic organ, not for breathing. The lobe finned fish perfected and modified the air sac for a breathing apparatus. An offshoot of this primitive ancestor of the amphibians is the lungfish, three species of which survive (Fig. 42). In addition to functional lungs, two of the living types have big fleshy lobes which enable them to walk. Devonian types were similarly constructed. By the time of the Late Devonian and Early Mississippian, lobe finned

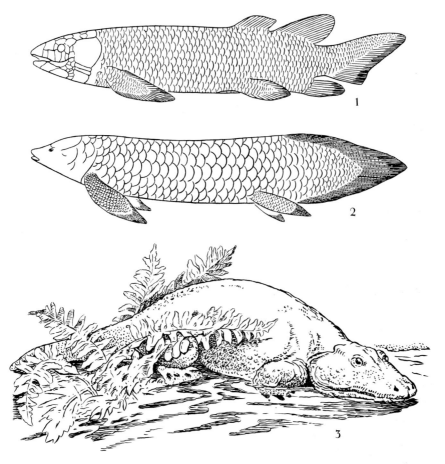

Figure 42. Lungfishes and amphibians. 1. *Dipterus,* Devonian lungfish (approximately X1/4); 2. *Epiceratodus,* Australian lungfish living today, some up to 5 feet in length; 3. *Eryops,* one of the largest amphibians reached 6 feet or so in length, Permian, Texas. (After Colbert.)

fishes were so well adapted to spending part of their life out of water that they are called amphibians.

Amphibians

From an evolutionary viewpoint, the amphibians (Fig. 42.3) can be regarded as important principally as the transition between the fishes and the reptiles. They were never as abundant nor have they been as geologically useful as their ancestors or descendants.

The amphibians are important because they were the first group of semiterrestrial vertebrates (Fig. 42). They are discussed here rather than with the terrestrial vertebrates because they were aquatic, primarily, and have never really solved the problems of life on land. They have been more successful in evolving an adequate air breathing apparatus than in solving the problem of skin drying or of maneuvering their bulk under the full force of gravity with no water buoyancy to help. Reproduction has always been tied to water. Paleozoic amphibians were forced to return to the water to lay their eggs in the protective and nourishing aquatic environment, and amphibians living today still reproduce in this manner.

Swimming Reptiles

Although the transition from amphibians to reptiles was one of the attainment of a completely terrestrial habitat, some reptiles retained an amphibious mode of life while others returned to the water and evolved structures for a successful aquatic life. A few modern reptiles still are aquatic (turtles, crocodiles, and some snakes), but the most remarkable adaptation to the aquatic life occurred in the Mesozoic. Ichthyosaurs were porpoise-like reptiles streamlined for fast swimming and fish and squid eating and ranged in size from 1 foot to more than 10 feet (Fig. 43.2). These swimming reptiles had long slender skulls, an efficient set of fish-catching teeth, specialized eyes, and well-developed paddles for guiding through the water. Hundreds of well-preserved specimens have been excavated from Mesozoic rocks in Germany and the United

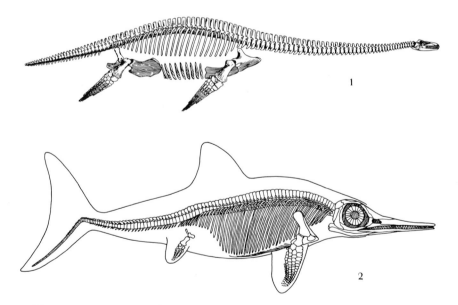

Figure 43. Swimming reptiles. 1. *Muraenosaurus,* 21 foot Jurassic plesiosaur; 2. *Ichthyosaurus,* 4–10 foot Mesozoic type, here showing body outline around skeleton. (After Romer.)

States. Specimens from Germany have delicate impressions of the skin preserved with the bony skeleton.

Other marine types include the plesiosaurs (Fig. 43.1) with long necks, short tails, and stout swimming paddles; the mosasaurs, which have short necks and long tails and are related to modern lizards; and other kinds of "sea monsters" which could justifiably serve as models for the sea monsters of legends. Most were short-lived and extinct by the end of the Cretaceous. Their ecologic replacements in modern seas are the dolphins, porpoises, and larger whales.

Swimming Birds

Aquatic birds include penguins as well as a large variety of diving birds, loons, and ducks. The adaptation to an aquatic environment was achieved early in the history of flying animals, and there is debate among specialists as to whether aquatic flightless forms evolved from flying types or not. A poor geologic record has not provided adequate data to answer the question.

Swimming Mammals

Mammals evolved in the Triassic, and over a period of 150 million years gave rise to several aquatic groups. Among the most successful are the cetaceans—the whales and their kin (Fig. 44). Whales evolved in the Early Cenozoic perhaps from a carnivore ancestor. This successful adaptation to an aquatic environment was accomplished relatively quickly and the evolution of the cetaceans is such that the earliest fossil forms known had already become excellently modified for swimming. Lungs are maintained and some forms can stay under water with a single breath for an hour.

The trend toward giantism is especially well illustrated by whales. The earliest Cenozoic varieties were more than 60 feet in length. The modern large blue and green whales are the largest animals that have ever lived. Some are 100 feet or more in length, which is several feet longer than the greatest dinosaur, and may weigh 150 tons—several times heavier than the greatest Mesozoic dinosaurs.

Even though fossil cetaceans are not common, sufficient material has been found to show that the small- and medium-sized types (such as porpoises, dolphins, killer whales, and narwhals) as well as the large toothless whalebone types, evolved from a common ancestor.

Although the cetaceans are the single most important group of aquatic mammals, several other mammal groups include aquatic members. These other groups are the carnivores, one branch of which consists of the pinnipeds (the seals, sea lions, and walruses). Swimming by limb motion characterizes the pinnipeds, and their greatly modified appendages as well as dentition have been distinctive since the Eocene.

Figure 44. Whales. 1. the Eocene ancestral whale, *Zeuglodon;* 2. a dolphin; 3. modern sperm whale; and 4. modern blue whale. (All approximately ✕1/20.) (After Colbert.)

The sirenians, including sea cows (Fig. 45) and dugongs, evolved during the Eocene and seem to be most closely related to the elephant family. The earliest fossil representatives show a dentition pattern similar to other hoofed mammals. Progressive modification of the pelvis structure in fossil and living types shows clear indication of evolution from a tetrapod (four-legged) ancestor. A vestigal pelvis is present in modern dugongs, and the fossil record of these "mermaids" can be related to Oligocene ancestors.

The rodents are a successful group of mammals whose members

Figure 45. Sea cow, *Manatus,*
marine member of a group that includes the elephants and kin (Eocene to Recent). (From photograph
of original specimen that was 8
feet long.)

include aquatic types such as the beaver and the martin. Among hoofed
animals, the hippopotamus might be considered an aquatic member of
the artiodactyl group. This group is agile on the land as well and its
adaptation to aquatic life may be a trend which is only in its beginning
stage.

FLIGHT IN VERTEBRATES

Among vertebrates, flight has been achieved by reptiles, birds and
mammals. The so-called flying fishes are an interesting semiaerial
group.

Reptiles

The multiple problems of flight were at least partially solved by the
reptiles during the Jurassic. Several kinds of flying types evolved and
were somewhat successful for a relatively brief period, becoming extinct during the Cretaceous. Flying organisms seldom leave good fossil
records and our knowledge of flying reptiles is based on few specimens
(Fig. 46). Most of the flying reptiles evidently lived in an area close to
a marine environment where fish catching was a mode of life. Most
were only a foot or so in length, but some kinds with a wing span of 50
feet have recently been found. These were the largest organisms to fly.
Footprints, suspected to have been made by flying reptiles, have been
found and from these it appears that they were extremely awkward on
the ground. Questions of skin covering, metabolism rate, and flying
capabilities with less than satisfactory wing structures, are unsolved
problems of the flying reptiles.

Birds

The first birds appeared in the Jurassic and marvelous details of
their feathers have been preserved, permitting the differentiation of

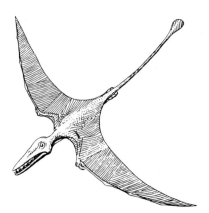

Figure 46. Flying reptile, *Rhamphorhynchus*, a Jurassic form, approximately 2 feet long. Other members of the flying reptiles had wing spans of 50 feet and are the largest flying organisms. (After Colbert.)

these birds from the flying reptiles. These two flying groups probably evolved from the same reptilian stock. The Jurassic and Cretaceous types of birds had teeth, and although the geologic record is poor, it clearly indicates that this was a primitive condition lost in advanced Cenozoic forms. The anatomy of birds has been quite uniform since the beginning of the Cenozoic, and external nonpreservable characteristics are used in differentiating living species. Very few complete bird skeletons (outside of aquatic types) have been found, and although knowledge of living species is far advanced, bird paleontology is less adequately understood. The great similarity of the earliest birds to certain small dinosaurs suggests a dinosaurian ancestry. If this is correct, birds are the closest living relatives of the dinosaurs.

Mammals

A high degree of specialization for flight was attained in the Early Cenozoic by the chiropterans, which include the bats (Fig. 47). Flying blind with the aid of a sonar-type navigation system, the bats are very proficient in flight. The earliest fossil bats from the Eocene are quite well advanced, and our knowledge of ancestors is lacking.

Figure 47. Flying mammal, *Myotis,* the little brown bat (X3/4). (After Young.)

Semiflying mammals include the flying lemur, an insectivore that glides using a large membrane that extends from the neck to the legs and tail. Fossils from the earliest Cenozoic may represent ancestral forms. The flying squirrel is a rodent proficient at gliding from tree to tree and represents another group that is successful at semiflight.

TERRESTRIAL VERTEBRATES — THE REPTILES

The amphibians, discussed with aquatic vertebrates, are important primarily because they were the first vertebrates to adapt, at least in part, to a terrestrial existence. The changes that took place during transition between amphibians and reptiles is so complete that it is difficult to characterize differences between an advanced reptile-like amphibian and a primitive amphibian-like reptile. The transition was made during the Pennsylvanian, and by the Permian reptiles were successfully competing with their amphibian ancestors. Perhaps the most important difference between these primitive reptiles and their amphibian ancestors was the ability of the reptiles to lay an amniote egg. This egg, with a durable calcareous covering and a rich glob of enclosed nutrient, freed the vertebrates from their cyclic return to water. The reproductive process was now completed on land. There are other preservable morphologic differences of importance including modifications of the skull, backbone, shoulder bones, ribs, and feet.

The earliest reptiles ranged from approximately 1 to 8 feet in length. Some were plant eaters, but a few were equipped with a complement of teeth, which suggests definite carnivorous habits.

Reptiles evolved rapidly in the terrestrial environments, and during the Permian and Triassic they were the most abundant and widespread land vertebrates. Several different reptilian groups acquired mammal-like characters during this interval, and it is thought that different primitive mammals had different reptile ancestors.

Among the most interesting aspects of reptilian evolution, three facts stand out: (a) a trend toward large size, even giantism, in many groups; (b) a trend toward adaptation in diverse environmental situations (including aquatic and aerial adaptations, previously discussed); and (c) massive extinction so that only a few of the dozen or so different Mesozoic groups survive to the present. There are many other factors, but these three have special interest.

Trend Toward Large Size

Within the limitations of the genetic makeup of a population, growth in reptiles is not limited as in mammals, and most reptiles continue to increase in size until death (Fig. 48).

Beyond this ontogenetic characteristic, most reptiles also evolved members of very large size, out of proportion to that shown by other

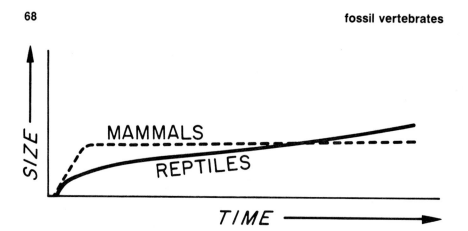

Figure 48. Diagrammatic plot of growth of reptiles and mammals. Mammals are more or less stable after sexual maturity but growth in reptiles continues throughout life.

vertebrates. This evolution is well illustrated by the dinosaurs, an important group whose ancestors were the same as those for flying reptiles, birds, and crocodiles.

There were two large groups of dinosaurs, each characterized by a particular pelvic structure (Fig. 49). One, the saurischians, had a triradiate structure with pubis bone extending down and forward beneath the ilium and adjacent to the ischium which extends backward and downward. The second group, the ornithischians, was characterized by a tetraradiate pelvic structure, so named because of a forward

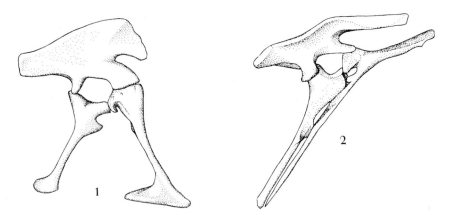

Figure 49. Pelvis structure in dinosaurs. 1. saurischian triradiate type, pubis extending forward and down, ischium extending backward and down, under the ilium; 2. ornithischian tetraradiate pelvis, pubis extending upward and backward, ischium in same position as in saurischian, and ilium with forward component of growth above. (After Romer.)

projecting pubis which also grew downward for another component of growth. The ilium and ischium had similar dimensions to those of the saurischians (Fig. 49).

Saurischians

There are two varieties of saurischians: theropods, the only flesh-eating dinosaurs, such as *Tyrannosaurus* and *Allosaurus;* and sauropods, the largest land animals that have ever lived. (Fig. 50). Some sauropods were 80 to 90 feet in length and weighed 40 to 60 tons. They were vegetarians and primarily quadripedal. Among the largest of the sauropods were *Diplodocus* and *Brontosaurus.* These were true giants and the largest tetrapods that have lived. A smaller member of this group of giants is shown on Figure 50.

The question might be asked as to the advantage of large size, since without a mammal-like metabolism to control body temperature, reptiles are at the mercy of their external environment. An interesting

Figure 50. Saurischian dinosaurs. 1. *Allosaurus,* a 40-foot theropod of the Jurassic; 2. *Ornithomimus,* a six-foot dinosaur of the Cretaceous; 3. *Camarasaurus,* a 20-foot sauropod of the Jurassic. (1, 2, after Colbert; 3, after Romer.)

experiment a number of years ago by the dinosaur specialist E. H. Colbert answered this question, in part. This investigator determined that the larger the reptile, the smaller the total body heat fluctuation in heating and cooling situations. This suggests that larger bodies can tolerate a wider fluctuation in cooling and heating (Fig. 51). The significance of this is modified by recent studies that suggest the dinosaurs, unlike most other reptiles, were "warm-blooded." Another advantage in size could be the factor of protection against predation, important in an age of active predators. But size also has disadvantages. These range from the inability to find shelter from severe climatic conditions to the danger of accidents. (A broken bone often is fatal for a large animal, such as an elephant, but not for smaller mammals.) One other consideration is the neurological fact that even the 1/10 of a second delay which might occur in sending nerve impulses along a 100-foot track from head to tail could prove fatal. In spite of these disadvantages, selection for large size seems to have been greater than selection against it, and many large forms evolved.

Ornithischians

The ornithischians were a more diverse group in their evolution but none were as large or ferocious as some of their saurischian cousins. Four distinct varieties of ornithischians evolved during the Jurassic and Cretaceous and then became extinct. Their evolution was, in part, due to selective pressures of their carnivorous sauropod cousins. Their extinction has been used to mark the close of the Mesozoic.

Ornithopods: This was the primitive group and includes the duck-billed dinosaur and its kin (Fig. 52.3–4). Most types were bipedal and

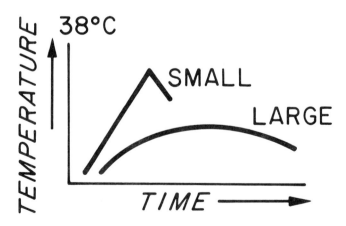

Figure 51. Relationship of body temperature fluctuations for large and small size reptiles. Smaller bodies gain and lose heat more rapidly than large bodies. (From Colbert and others.)

Figure 52. Ornithischian dinosaurs. 1. *Stegosaurus,* a 20-foot Jurassic vegetarian with bony plates on its back; 2. *Ankylosaurus,* also 20 feet or so in length with heavy body armor, Cretaceous; 3. *Pachycephalosaurus,* an ornithopod with a thick bony growth above the brain; 4. *Corythosaurus,* a 30-foot, semi-aquatic ornithopod with a skull modified for nasal special-izations such as snorting, a possible sex-linked character. (After Colbert.)

some had a remarkable skull modification including nasal passages which may have permitted snorting sounds used in attracting mates, and probably also increased their sense of smell. Other modifications appeared, the functions of which escape us today. One type had more than 6 inches of solid bone for a skull battering ram (Fig. 52.3).

Stegosaurs: These dinosaurs were characterized by the genus *Stegosaurus,* a 20-foot quadrapedal form with a small head and a large body with a row or rows of long bony plates along the dorsal midline (Fig. 52.1). These were 20 feet in length and, in one way, the least successful of the dinosaurs: they appeared during the Early Jurassic and became extinct early in the Cretaceous, the shortest geologic range of any of the major dinosaur groups. Also, they are one of the smallest groups, only two families and less than a dozen genera known.

Ankylosaurs: These were the armored dinosaurs, the Cretaceous successors of the stegosaurs. About the same size as the stegosaurs, they had bony plates which must have covered most of the body (Fig. 52.2). Nearly twice as many types of ankylosaurs are known as stegosaurs.

Ceratopsians: The last major group of dinosaurs to appear are the Cretaceous *Triceratops* (Fig. 53.2), and another dozen or so horned types. Development of a skull with horns and a neckskirt was a trademark of this group. The largest types were 25 feet long and weighed six or more tons. Great herds of ceratopsians ranged over North Amer-

Figure 53. Ceratopsian dinosaurs. 1. *Protoceratops,* ancestral ceratopsian and its eggs, known from abundant material found in Mongolia, about 9 feet in length; 2. *Triceratops,* 25-foot Cretaceous dinosaur with well developed horns. (After Colbert.)

ica and numerous fossils have been collected. One dwarfed group, including eggs, has been discovered in Mongolia (Fig. 53.1). It appears that the ceratopsians were still expanding and numerous, when, for reasons still unknown, they, along with the other dinosaurs and many other types of reptiles, became extinct.

The trend toward large size was prominent among other reptiles, as well. Mammal-like reptiles with large fins, as well as turtles, snakes and crocodiles, all had large members (Fig. 54).

Adaptation to Diverse Environments

Some aspects of reptilian adaptation have been discussed under aquatic and aerial headings. Terrestrial adaptations include predation, browsing, burrowing, adaptation to semiaquatic and arboreal environments, and most of the other terrestrial adaptations seen among modern mammals.

Extinction

The modern turtles, snakes, lizards, and crocodiles, essentially are the only groups that have survived of a lineage that included numerous reptilian categories during the Mesozoic. The extinction of the various

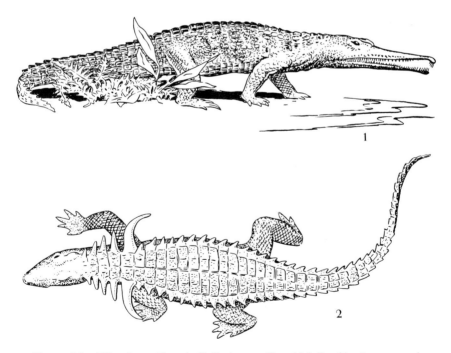

Figure 54. Triassic reptiles. 1. *Rutiodon,* reptile which lived in streams and lakes; 2. *Desmatosuchus,* an armored form with heavy plates and spines. Both attained lengths of 10 feet or more. (After Colbert.)

types of reptiles was progressive throughout the Mesozoic but was more dramatic late in the Mesozoic. Several kinds of dinosaurs, flying reptiles, and swimming reptiles became extinct within a relatively short period of time. Major climatic change, disease, continental drift, competition with the mammals, change in food sources, and various kinds of catastrophes, have been suggested as explanations for their extinction, but no single item appears possible to explain such a significant event. A combination of things may be the answer. One investigator reported recently (Erben, 1970) that many fossil reptile eggs from the Cretaceous have a shell thickening similar to that shown by modern chicken eggs with calcification problems. The thickening of an egg shell restricts the transfer of oxygen and carbon dioxide and adversely affects the embryo. This, plus a combination of other events such as those enumerated above could have been the factors to which the reptiles were unable to adapt. And inability to adapt is, in the final analysis, the primary reason for extinction.

TERRESTRIAL VERTEBRATES —THE MAMMALS

When the reptiles were at the height of their evolutionary success, the mammals appeared. Their ancestors were among the mammal-like reptiles of the Triassic. In fact, several ancestors have been identified and it appears that different primitive mammals had different reptilian ancestors. The oldest fossil mammals are known from teeth and jaws, a few bones, and little more. The teeth are distinctive, however, and by themselves give a good indication of size, age, diet, and indirectly, the environment in which the animal which bore them lived. For example, bladelike teeth indicate a carnivorous diet, low-crowned teeth with broad surface areas indicate browsing tendencies, and high-crowned teeth with broad surfaces indicate grazing habits. Combinations of these types indicate different habits, e.g., bladelike plus broad surface types indicate omnivorous habits. Of all the skeletal remains, teeth are by far the most important and instructive in the study of mammals. Unlike their reptilian ancestors, mammals had highly differentiated teeth, that is, incisors, canines, premolars, and molars—a specialization which had no little role in their success. The skull and postcranial skeleton are less important because most mammals, while differing in proportion as well as in certain details, have a remarkably uniform skeleton, exclusive of the teeth.

There are, of course, many other differences between the mammals and the reptiles. Features of the brain, skin, ear structure, nasal structure, pelvis, jaw, and body temperature control are only a few of these differences. Many of these features have not been preserved as fossils but the sturdy calcium phosphate teeth are readily preserved and have given most of the data that we have concerning many mammals.

Mammals are more complex organisms than reptiles. They show some of the same trends, but differ in many significant ways. First, mammals became adapted to more diverse environments. In addition, while the whale, elephant, and rhinoceros suggest some giantism, most successful mammals are moderate- to small-sized forms that are distinguished by specialized brains and teeth. There was great development of that portion of the brain that controls the reasoning processes. In addition, the specialization of teeth that have become adapted for almost every variety of diet has been a very positive factor in the success of mammals.

One problem in understanding the evolution of mammals has been their present distribution in terms of the present geography of the world. Why are there marsupials in Australia and the Western Hemisphere (both living and fossil), but not elsewhere? Why are the fossil hoofed forms in South America unlike those elsewhere (including North America)? Why are there such obvious kinships among modern terrestrial mammals now separated by oceans?

In spite of known mobility of mammals (island hopping, etc.), these are difficult questions to answer. However, recent theories that picture the Earth's surface as a series of moving crustal plates, including mobile continents and oceans, have provided an answer. We now understand that there probably has been a progressive breakup of the continents, beginning 180 million years ago—a time that quite neatly coincides with mammalian evolution. The progressive shift of continents through time explains past and present distribution without calling on the hypothetical land bridges that before seemed necessary (Fig. 55).

Primitive Types and the Marsupials

Mesozoic mammals were small and rodent-sized. Several tooth types have been described. Both the marsupials and their close relatives, the placentals, evolved from one of these types during the Late Mesozoic. The marsupials (Fig. 56) are more primitive than placentals in many respects but along with the aberrant monotremes (duck-billed platypus, spiny anteaters) have nevertheless survived to the present.

The centers of marsupial evolution were in Australia and South America, parts of a former single continent, where, because of geographic isolation, these animals were able to gain a foothold before their placental cousins. Competition with placental types hindered their development in South America, but the separation of Australia isolated the marsupials and few placentals reached Australia before man brought them there. In this secluded environment, marsupials evolved much the same way as the placentals did in other parts of the world. Marsupial animals, including wolves, cats, rabbits, squirrels, and bears have evolved in Australia during the Cenozoic, and in many ways are much like their placental counterparts which evolved on other conti-

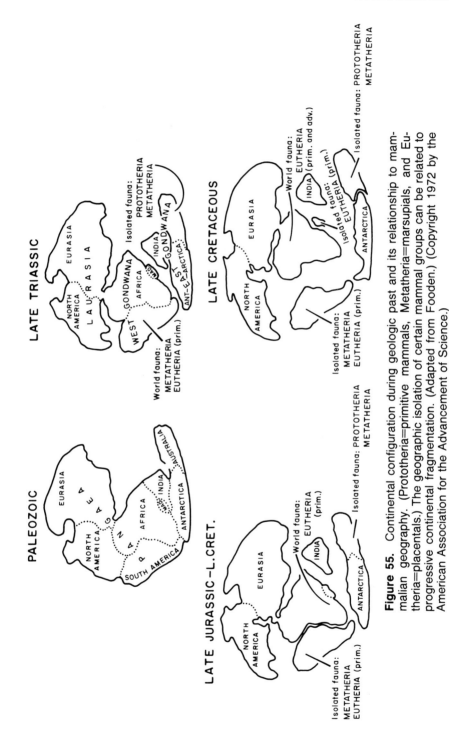

Figure 55. Continental configuration during geologic past and its relationship to mammalian geography. (Prototheria=primitive mammals, Metatheria=marsupials, and Eutheria=placentals.) The geographic isolation of certain mammal groups can be related to progressive continental fragmentation. (Adapted from Fooden.) (Copyright 1972 by the American Association for the Advancement of Science.)

Figure 56. A marsupial, the American opossum, *Didelphis*. This is a living representative of the marsupial group in the western hemisphere. Geologic range is Cretaceous to Recent. Original about 18 inches. (Adapted from Young.)

nents. This is an excellent example of how similar environmental conditions can encourage similar morphologic adaptions. Such parallel evolution can be noted in other fossil groups, but perhaps no other group illustrates the principle so well (Clemens, 1968).

Placental Mammals—A Success Story

Approximately 95 percent of the mammals which have lived during the Cenozoic are placentals. The placentals moved into the environmental niches vacated by the Mesozoic reptiles and filled these vacancies in a much more successful way than they had ever been filled before. Almost anything that the reptiles did—swim, fly, or bite each other—was done bigger and better by mammals. Flying mammals (bats) are better flyers than were the pterodactyls. Swimming mammals (whales and kin) are at least as good if not better swimmers than the ichthyosaurs. On land, mammals have developed more diverse habitats than their reptilian progenitors. Cenozoic mammals are so diverse and have left such a good fossil record that their study is worthy of a separate and more exhaustive treatment than can be given here. The classification of more than 2,600 fossil and living genera (with many times that number of species) is complex and most specialists would recognize at least 28 different orders. More than half this number of orders are living today. The advanced terrestrial placentals include the insectivores (moles, hedgehogs, shrews), anteaters, sloths, primates, the large variety of larger hoofed types, and several extinct varieties.

Primitive Placentals

Some of the more interesting are the large ground sloth and the glyptodont (which is an extinct relative of the early insectivores) and

Figure 57. Placental mammals. 1. Giant ground sloth, some more than 12 feet high, Late Cenozoic; 2. heavily armored glyptodont, 6 feet or so in length. (After Colbert.)

the primates (Fig. 57). As is the case with most arboreal organisms, the geologic record of primates is poor. They apparently evolved from the tree-shrew and lemur groups. Higher types, including monkeys, apes, and man, probably had a common biologic ancestor during the Late Cenozoic. Some students of primate evolution maintain that the record of man's evolution from the Pliocene to the Recent is now rather complete, and that missing links may no longer be stylish (Fig. 58).

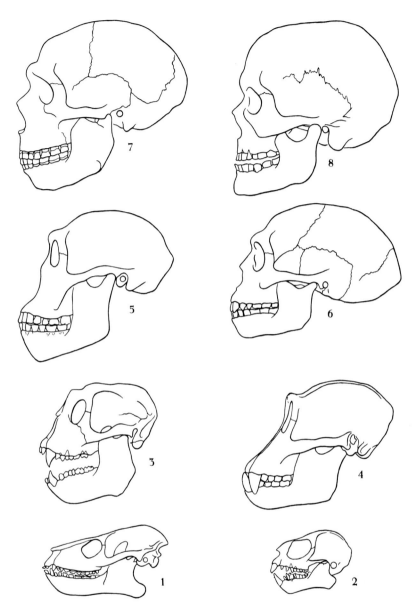

Figure 58. Primate skulls. 1. *Notharctus,* an Eocene lemur; 2. *Tetonius,* an Eocene tarsioid; 3. *Mesopithecus,* a Pliocene monkey-like primate; 4. *Pan,* modern chimpanzee; 5. *Australopithecus,* Pleistocene man-like primate; 6. *Homo erectus,* primitive Pleistocene man; 7. *Homo neanderthalensis,* Late Pleistocene man; 8. *Homo sapiens,* Cro-magnon man. (1–3, approximately ×1/2; remainder approximately ×1/4. Arrangement represents various morphologic changes.) (After Colbert.)

The rates of evolution for man and his ancestors are high—much higher than the one species per million years which has been suggested as a reasonable rate for certain other mammals. The growth of the brain and the development of societies are only two of the many factors which are responsible for this unique event in mammalian evolution.

Hoofed Placentals

The large and diverse group referred to as hoofed placentals actually includes a large number of distinctly different living and extinct kinds. Included in this group are the even- and odd-toed mammals (cattle, horses, and kin), and the elephants. Extinct types include forms quite unlike any living today. *Thomashuxleya,* an Eocene sheep-sized animal, was an early member of one group whose descendants included the Pleistocene *Toxodon,* a massive animal that stood 6 feet high at the shoulders (Fig. 59). Charles Darwin discovered the first specimen of this extinct species in South America. These ancient hoofed animals included a wide range of rodent- to rhinoceros-sized forms. One large group was restricted to South America and became extinct in the late Cenozoic.

Two of the most important hoofed types are the odd-toed (perissodactyl) and even-toed (artiodactyl) groups. These two groups evolved rapidly during the Cenozoic. The perissodactyls have passed the height of their development and are near extinction. The even-toed group are probably at the height of their development today.

The odd-toed group includes the extinct titanotheres and chalicotheres, the rhinoceros, and the horses and tapirs. Titanotheres were 7- to 8-foot animals with large horns and their chalicothere cousins were unique in possessing claws for digging. Tapirs of South America and Asia are living representatives of this primitive group. The rhinoceros and horse groups were much more diverse formerly and both seem headed toward extinction.

The evolution of the horse has been described for years as an almost ideal example of evolution. Views presented have sometimes oversimplified a very complex evolutionary history, and it has been pointed out that the famous and often-cited reduction in number of toes is only one factor drawn from many similar factors, but from many different horse groups. Increase in size, complication of tooth structure, and skull modification are but a few of the other important changes which have affected various groups of horses during the last 60 million years (Fig. 60). All of these taken together must be understood for the real story of horse evolution.

The even-toed mammals are very successful hoofed types of the late Pleistocene and Recent. They probably are as widespread and

Figure 59. Ancient hoofed types. 1. *Thomashuxleya,* a 5-foot Eocene South American mammal; 2. *Toxodon,* Pleistocene rhinoceros-sized mammal. (After Colbert.)

numerous today as they have ever been. Pigs and hippopotamuses, camels and their kin, cud-chewing cattle, sheep, and deer are only a few of the numerous kinds of living representatives of these mammals familiar to everyone. The camels first appeared during the Early Cenozoic in North America (Fig. 61). Later, they migrated to the Eurasian area and became extinct in North America, but today are represented in South America by the llamas. The llamas have continued with only modest change since the Miocene.

Elephants have always been among the largest hoofed mammals. Little is known of the early evolutionary history of this group of animals which first appeared in the Eocene. Good specimens found in the Miocene show that the trend toward giantism was well established early.

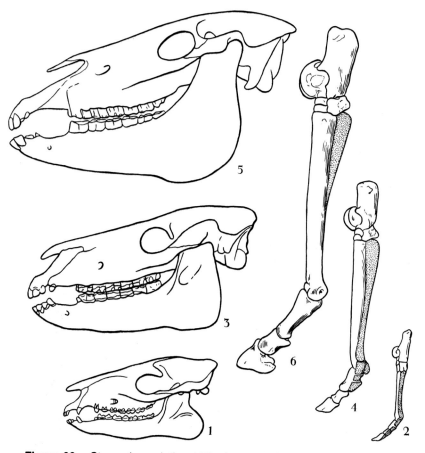

Figure 60.   Stages in evolution of the horse. 1, 2. skull and hind foot of primitive horse, *Hyracotherium,* Eocene (approximately X1/4); 3, 4. skull of *Parahippus* and hind foot of *Merychippus,* Miocene (approximately X 1/8); 5, 6. skull and hind foot of modern horse, *Equus* (approximately X 1/4). (After Colbert.)

The Pleistocene could be called the Age of Elephants. During this time several varieties of now extinct types including the wooly mammoth (Fig. 62.2) spread over much of North America. Their cousins, the mastodons, were also widely scattered in North America at this time.

Wooly mammoths and mastodons were contemporaneous with early man. Mastodons probably were wandering in North America as recently as 8,000 years ago and legend says they may have existed until a few hundred years ago. In the Alaskan gold fields, thawing of frozen soils has revealed a great amount of the skin, hair, and bone of these

Figure 61. Artiodactyls. The camels. 1. *Stenomylus,* small Miocene form approximately 2 feet high; and 2. *Alticamelus,* Pliocene camel, approximately 10 feet high. (After Colbert.)

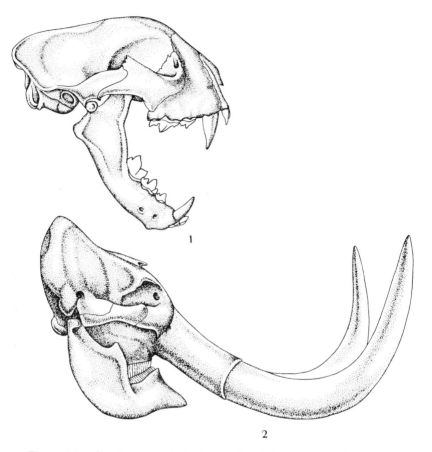

Figure 62. Carnivores and elephants. 1. a Pliocene cat, *Metailurus,* skull approximately 6 inches long; 2. *Mammuthus,* the wooly mammoth (X1/20). (After Romer.)

animals. Whole specimens, meat and all, have been recovered from ice in Siberia. Mastodons and mammoths became extinct only a short time ago because of overkill by man, according to some theories, and today only two elephant species, the Indian and African, remain.

Carnivore Placentals

The meat-eating mammals of the Cenozoic show excellent examples of adaption to a predatory mode of life. Their whole anatomy, including the jaw and tooth structure, illustrates complete adaption for carnivorous living. Early forms showed some of these same features and during the Middle Cenozoic, a variety of minks, skunks, badgers, weasels, cats (Fig. 62.1), and dogs appeared and became well adapted to the ecologic niches which they occupy today.

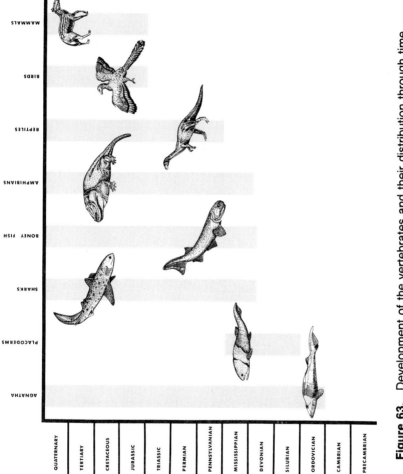

Figure 63. Development of the vertebrates and their distribution through time.

REFERENCES

Clemens, W. A. 1968. Origin and early evolution of marsupials. *Evolution* 22:1–18.

Colbert, E. H.; Cowles, R. B.; and Bogart, C. M. 1946. Temperature tolerances in the American alligator and their bearing on the habits, evolution, and extinction of the dinosaur. *American Museum of Natural History Bulletin* 86:327–74.

Colbert, E. H. 1969. *Evolution of the vertebrates.* 2d ed. New York: John Wiley and Sons.

Erben, H. K. 1970. Dinosaurier; Pathologische Strukturen der Eischale als Letalfaktor. *Umschau* 69:552–53.

Moy-Thomas, J. A. 1971. *Palaeozoic fishes.* 2d ed., revised by S. Miles. Philadelphia: W. B. Saunders.

Romer, A. R. 1966. *Vertebrate paleontology.* 3rd ed. Chicago: University of Chicago Press.

Savage, D. E. 1975. Cenozoic—the primate episode. In *Approaches to primate paleobiology,* ed. H. Kuhn et al. *Contrib. Primate,* vol. 5, pp. 2–27. Basel, Switzerland: Karger.

Stahl, B. J. 1974. *Problems in evolution.* New York: McGraw-Hill.

Young, J. Z. 1950. *The life of vertebrates.* Toronto: Oxford Press.

Chapter Six

Fossils as Tools in the Interpretation of Earth History

To this point we have reviewed the appearance of life, and its evolution and diversification, as well as the problems related to the preservation and interpretation of its fossil remains.

We turn our attention now to how fossils can be used in problem solving. Specifically, our interest will be the use of fossils in solving the kinds of problems for which they are uniquely suited. Problem solving with fossils falls into at least four categories: (a) fossils in biological interpretations, (b) fossils in evolution and biostratigraphy, (c) fossils in paleoecology, and (d) fossils in paleogeophysics. The first of these will be discussed in this chapter. A separate chapter will be devoted to each of the other three.

FOSSILS IN BIOLOGICAL INTERPRETATIONS

The various structures that have evolved with any organism are adaptations for one or more functions. The kinds of structures and their functions include a variety of things such as tetrapod limbs, bird wings, and fish fins (as aids in locomotion) and teeth (as aids in feeding). These things are obvious. These things are readily apparent and therefore the relationship between structure and function is easily understood. This is true particularly for vertebrates. For example, the function of the spear-like canine of a sabre-tooth cat (Fig. 64) with a jaw constructed to open 90°, can be surmised easily even though the animal is extinct.

But what was the function of elaborate spines on an extinct clam or the ribs on a cephalopod? The study of fossils with the aim of complete biological interpretation of structure and function is a challenging part of paleontology. We can assume that every structure that has evolved had one or more functions. This is a generalization and probably there are exceptions. Most likely, however, the exceptions are rare.

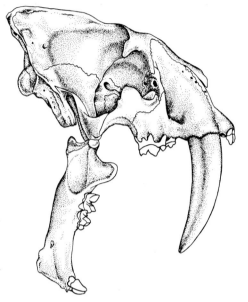

Figure 64. Pleistocene Sabre-tooth cat, *Smilodon.* Notice extent of jaw opening and powerful upper canine. Original skull length approximately 1 foot. (After Romer.)

Functional Morphology

A splendid illustration of biological interpretation has been given by Carter (1972) who described several species of clams from the Cretaceous chalk of Western Europe. These forms have an array of spines on one valve, and an additional morphologic feature is the deep spoon shape of the valve (Fig. 65). If we assume that these structures were adapted to some function, how is that function determined? Carter reasoned that the soft chalk ooze that was the animal's substrate in life gave little support to a normal bivalve. As the bivalve grew and its weight gradually increased relative to its shell area, there must have been a tendency for the shell to sink into the ooze. His investigations indicated that the spines on the lower valve were regularly spaced on the shell flanks and had the shape of a flattened blade. They could therefore have functioned as "snowshoes" for support in the soft ooze substrate. The deep spoon shape of the lower valve was probably an additional adaptation to this same problem as it permitted the individual to settle into the ooze somewhat but become stable before it was buried.

A different kind of biologic interpretation was presented by Seilacher (1968). This investigator studied the problem of interpreting the morphology and life habits of a group of cephalopods that have been extinct for millions of years. He demonstrated that a calcareous rostrum (Fig. 66), a structural feature of this extinct group of cephalopods, probably was borne externally during life. This was based on the observation

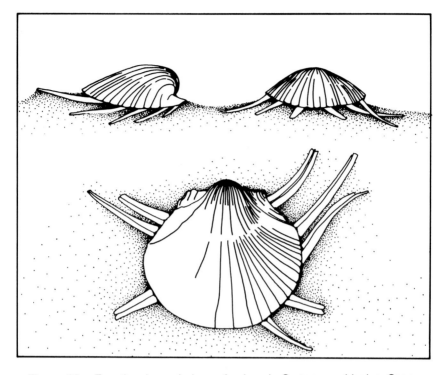

Figure 65. Functional morphology of spines in Cretaceous bivalve *Spondylus spinosus*. This clam was adapted to very soft substrata by development of broad spines at right angles to shell margin. These "snowshoes" gave the clam a degree of stability in the soft chalk substrata of the Cretaceous seas. (Modified from illustrations by Carter.)

that there were cirriped (barnacle) borings in numerous specimens that had shapes and angles that could best be explained by being at or near the surface of the animal. It is known that certain kinds of small barnacles always bore so as to be situated facing the water currents. The shape of the borings on the extinct cephalopods demonstrated that the rostrum of the living animal was situated so that little or no fleshy covering was possible. Thus, understanding the living habits of barnacles enabled the investigator to substantiate that this ancient cephalopod's rostrum was external, at least during the later stage of the animal's life. In addition, the direction of movement of the extinct cephalopods must have been in line with the barnacle borings, a fact that was not known previously.

An investigation of jaw mechanics of certain dinosaurs has given similar kinds of data. Ostrom (1964) studied ceratopsian dinosaurs and particularly their jaw structures. A detailed analysis of muscle sizes

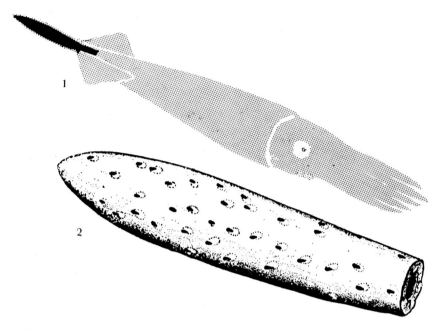

Figure 66. Belemnite. 1. reconstruction of the squid-like animal in swimming position with rostrum or guard (approximately 6 inches in length) shown
in black; 2. rostrum or guard with barnacle borings showing orientation as
in living animal. Representative of a group that ranged from the Devonian
to the Eocene. (Adapted from Seilacher.)

(from attachment scars on the bone), the leverage and fulcrum for
movement of the jaw, and similar matters, suggested that the animal
could transmit powerful shearing forces to the jaw and teeth. In fact,
it could transmit forces far greater than necessary for the passive herbi-
vorous habit that had been supposed for this animal. Ostrom's analysis
suggested that such function may have been an adaption for meat
eating, a completely different mode of life than previously suggested.

Thus, clam spines, barnacle borings, and dinosaur jaws can be interpreted in terms of function, and our understanding of the biological
operations of animals extinct for millions of years is enhanced.

Behavioral Patterns of Fossils Interpreted from Tracks and Trails

Behavior manifests itself in movement of some kind, the most
important of which is locomotion. Locomotor movement is of particular
importance in geology because it can affect and modify sediment and
be preserved as a record of fossil behavior. The resulting tracks, trails,
and burrows can be studied in terms of specific behavioral patterns.
Thus, directed motion from point to point represents one behavioral

pattern, such as crawling in search for food; burrowing into soft sediment may represent another, such as digging a shelter. Resting tracks, directed motion trails, burrows of shelter, and burrows for feeding, are four common behavioral patterns that have modified sediment and are preserved as trace fossils.

The behavioral traits that lead to sediment modification have important meaning in biological interpretations. The resulting trace fossils are very useful tools for biological and ecological interpretations. For example, the principal ecologic reason for an animal to burrow for a shelter probably is related to protection. Because we know that in modern seas, most shelter burrows are in well-illuminated shallow water where protection from predators is important, we can safely assume that fossil burrows of the same type served the same purpose. This interpretation gives us data on the behavior and ecology of the fossils (Fig. 67).

Organisms have evolved and environments have changed, but living organisms always are more or less adjusted to their environments and certain types of environments always have existed. Thus, there have always been shorelines at the ocean-continent margin and there always have been abyssal depths, although the geography has changed and former shorelines may later become the sites of abyssal depths and vice versa. Similarly, for at least 600 million years there have been invertebrate marine organisms, and one of the most important behavioral characteristics of these organisms has been their feeding behavior. An important part of evolution has been the evolution of feeding patterns that are most efficient in a particular environment. Therefore different feeding patterns have evolved in each of the many environmental settings. Furthermore, regardless of what kind of animal is doing the feeding, organisms that live in similar environmental niches usually have similar feeding behavior.

For example, the abyssal depths are characterized by a scarcity of nutrients. The only source of nutrients is the water column above the deep ocean floor. The water column supports a whole chain of pelagic organisms that have evolved feeding patterns for removing nutrients from the water. Because of this, nutrients that reach the deep ocean floor are very limited and organisms that have adapted to life in the abyssal depths have also adapted a very efficient behavioral pattern to retrieve the few nutrients that are present. One feeding pattern for this environment is a very systematic and complex pattern that carefully avoids any overlap or criss-crossing of sediment that has already been grazed or mined (Fig. 67). The resulting sediment modification has produced trace fossils whose structure becomes a key to a particular behavior and ecology. Thus, even the traces of life can be used in interpreting Earth History.

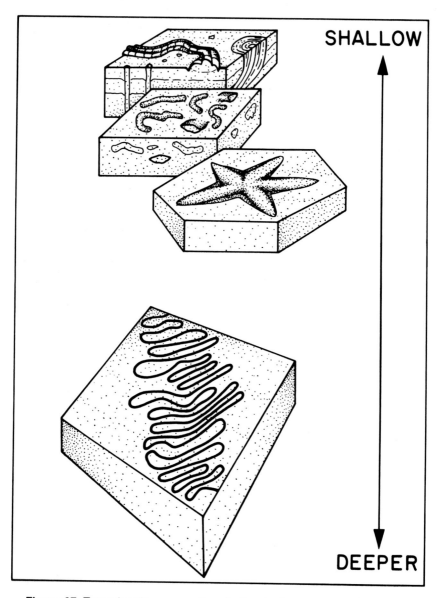

Figure 67. Trace fossils representing shallow to deep water environments. Shallow water types include vertical burrows, resting places, and random motion tracks. Deeper water forms are more systematic—the bottom example is a grazing trail suggesting a very systematic search for the limited nutrients in deeper water. (Modified from Chamberlain and Clark.)

Interpretations of fossil behavior have also been based on studies of modern organisms. For example, two oceanographers (Heezen and Hollister, 1971) have reported curious spirals (Fig. 68) at almost every station in the world's oceans (where photos are available) where the water depth was greater than 4000 meters. Bottom photographs were taken and they indicate that these spiral forms are largely absent at stations where water depth was less than 4000 meters. (The animals making the spirals where also photographed.) Similar coils found in ancient sediment thus allow interpretation of water depths that existed millions of years ago. Equally important is the fact that the behavior that produced the particular trace fossil becomes known.

Figure 68. Trace fossil *Spirophycus*, a fecal coil found in modern oceans at depths greater than 4000 meters. When it occurs in sedimentary rocks, the rocks are interpreted to be deep water deposits (X.5). (Adapted from Chamberlain and Clark.)

REFERENCES

Carter, R. M. 1972. Adaptations of British Chalk Bivalvia. *Journal of Paleontology* 46:325–40.

Crimes, T. P., and Harper, J. C., eds. 1970. *Trace fossils.* Liverpool: Seel House Press.

Frey, R. W., ed. 1975. *The study of trace fossils.* New York: Springer-Verlag.

Heezen, B. C., and Hollister, C. D. 1971. *The face of the deep.* New York: Oxford University Press.

Ostrom, J. A. 1964. A functional analysis of jaw mechanics in the dinosaur *Triceratops.* Peabody Museum of Natural History, *Postilla,* v. 88.

Schafer, W. 1972. *Ecology and palaeoecology of marine environments* (English version ed. by G. Y. Craig, tr. by I. Oertel). Chicago: University of Chicago Press.

Seilacher, A. 1967. Fossil behavior. *Scientific American* 217:72–80.

Seilacher, A. 1968. Swimming habits of belemnites—recorded by boring barnacles. *Palaeogeography, Palaeoclimatology and Palaeoecology* 4:279–85.

Chapter Seven

Fossils in Evolution and Biostratigraphy

The history of paleontology indicates that man first took fossils seriously when he discovered that they could be used for something. Fossils were first used as tools in biostratigraphy. As sequences of fossil groups became better understood, a relationship between evolution and the development of a system of biostratigraphy developed.

BIOSTRATIGRAPHY

In its simplest sense, biostratigraphy involves the application of fossils in establishing a relative time framework for sedimentary rocks. Thus, if fossil species A occurs in sediment that can be proven to be older than that in which species B occurs, and B occurs in sediment proven to be older than species C, the sequence of fossil species A, B, C, becomes a biostratigraphic sequence. The occurrence of A anywhere on Earth thereafter indicates that the sediment in which it occurs is older than that bearing B or C, and species A becomes an index fossil that can be used for widespread correlation.

Biostratigraphy developed in England where rocks first were differentiated into logical systems. The differences among the rock systems included lithologic differences, but final differentiation was based on the different fossils. In this way, Ordovician rocks were defined as being different from Cambrian and Silurian rocks because the fossils of the Ordovician were different.

To a certain extent fossils still are used most widely to classify and correlate sedimentary rocks sequences around the world. Biostratigraphy has been refined and refined again since the middle of the nineteenth century. Today, there is a worldwide workable and precise biostratigraphy for most unmetamorphosed sediment and this is based on a variety of key invertebrates, vertebrates and plants. The art of biostratigraphy is so refined that among some groups, the definition of zones and subzones approaches correlation (and classification) of time intervals of 1/2 million years or less. This is impressive in a science where synchronous events are those judged to have occurred in a 3 to 5 million year time spread.

For example, the Upper Devonian represents sediment deposited over an interval of approximately 15 million years. A biostratigraphy of 30 zones of conodonts divide and subdivide rocks of this series into units representing less than 1/2 million years. The zonation is based, in part, on ancestor and descendant evolution, as well as on unrelated species. The zonation can be used worldwide in identifying fine time intervals in Upper Devonian rocks (Fig. 69).

All fossils do not have equal value in biostratigraphy, however. Generally, pelagic organisms or other organisms that had opportunity for widespread distribution, that evolved rapidly and have easily recognized hard parts, make the best index fossils. Among the plants, spores and pollen meet these requirements. The evolution and distribution of pollen have particular value in the biostratigraphy of the late Cenozoic.

Among the primitive organisms, the protistans (one celled plant and animals), particularly the microscopic siliceous and calcareous forms such as nannoplankton, Foraminifera, flagellates, diatoms, and radiolarians, are of enormous value. In the exploration for petroleum, these kinds of fossils are widely used for the identification and correlation of rock layers. Larger microfossils such as conodonts and ostracodes are also extremely important biostratigraphic tools.

Pelagic graptolites and cephalopods are the best index fossils among larger invertebrates, but benthonic (bottom-dwelling) corals, brachiopods, bivalves, echinoderms, and trilobites have all contributed to a workable biostratigraphy.

Vertebrates have value, particularly in terrestrial or nonmarine sediment. Skulls of reptiles and teeth of mammals have use in Mesozoic and Cenozoic correlation. Microscopic fish teeth are abundant in Devonian and younger sediments and are also useful fossils.

Ammonoid cephalopods are among the most useful fossils in biostratigraphy. From the Devonian to the end of the Cretaceous (some 250 million years) the ammonoids evolved rapidly and left a good record that is used in various zonal schemes. Each of the seven geologic periods in this interval has been divided into ammonoid zones. The divisions for the Pennsylvanian and Permian are shown in Figure 70.

Evolution

The ability to use ammonoids in biostratigraphy is based on their rapid and at least partially understood evolution. Many of the successive zones represent relationships between ancestors and descendants. Of special interest in ammonoid evolution was the production of "heteromorph" ammonoids that do not coil planispirally (as is the case with

Protognathodus		30
unnamed zone		29
Spathognathodus costatus	UPPER	28
	MIDDLE	27
	LOWER	26
Polygnathus styriacus	UPPER	25
	MIDDLE	24
	LOWER	23
Scaphignathus velifer	UPPER	22
	MIDDLE	21
	LOWER	20
Palmatolepis quadrantinodosa	UPPER	19
	LOWER	18
Palmatolepis rhomboidea		17
Palmatolepis crepida	UPPER	16
	MIDDLE	15
	LOWER	14
Palmatolepis triangularis	UPPER	13
	MIDDLE	12
	LOWER	11
Palmatolepis gigas	UPPER	10
	MIDDLE	9
	LOWER	8
A. triangularis		7
Polygnathus asymmetricus	UPPER	6
	MIDDLE	5
	LOWER	4
	LOWERMOST	3
hermani cristatus	UPPER	2
	LOWER	1

Figure 69. Conodont zonation of the Upper Devonian. This detailed sequence of conodont zones represents a high degree of biostratigraphic refinement. Each division represents less than 1/2 million years. Parts of this sequence have been recognized in Upper Devonian rocks in most parts of the world.

Permian	Cyclolobus
	Timorites
	Waagenoceras
	Perrinites
	Properrinites
Pennsylvanian	Uddenites
	Prouddenites
	Eothalassoceras
	Wellerites
	Owenoceras
	Paralegoceras
	Gastrioceras

Figure 70. Ammonoid cephalopod zonation of the Pennsylvanian and Permian. The zonation of these two time intervals is based on genera that everywhere characterize the sediment deposited during these intervals.

most ammonoids) but are either uncoiled or coil helically or in some aberrant manner (Clark, 1965) (Fig. 71).

Heteromorph types evolved rapidly during the Mesozoic. Four genera appear in the Late Triassic, seven in the Jurassic, and several hundred during the Cretaceous. The aberrant coiling (or uncoiling) pattern of these forms has suggested degeneracy or racial senility to most workers. In addition, the fact that the time of greatest evolution of heteromorphs was in the Cretaceous coincident with the extinction of all ammonoids at the close of that period, suggested that the aberrant coiling habit was related to factors responsible for extinction. Indeed, most investigators were convinced that the heteromorph coiling was a symptom of the extinction that affected all ammonoids.

Detailed study of the evolution of heteromorphs suggested an alternate explanation to Wiedmann (1969). He carefully studied the growth stages of the heteromorphs, and more particularly the sequence of chamber wall changes and their evolution through time. His studies indicated that rather than "racial senility" as an explanation, the evolution of aberrant coiling patterns was more likely an adaptation to a more passive, perhaps benthonic mode of life, at least for Triassic and Jurassic heteromorphs. Some of the Cretaceous heteromorphs can be explained in terms of ecologic adaptation, as well. In addition, the

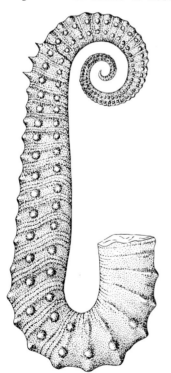

Figure 71. Cretaceous heteromorph ammonoid cephalopod, *Ancyloceras matheronianum.* Notice early loose coiling followed by straight shafts in later growth stages. This form probably was adapted to passive benthonic life (X.2). (Adapted from d'Orbigny.)

evolution of Cretaceous types shows an interesting pattern in many different families that may be related to an alteration in growth patterns caused by genetic changes. Thus, characteristics that appeared in the adult stages of ancestors may have begun to appear in the early stages of descendants. In this way, a normal adult characteristic such as a slight straightening of the coiling direction could be introduced in a descendant at an early stage and thus the entire shell would become abnormally coiled. Also, heteromorph types apparently gave rise to normally coiled descendants indicating that the development of aberrant coiling represented adaptations to environmental stimuli rather than symptoms of extinction (Fig. 72). The details of heteromorph development are a splendid example of evolution. The various shell types are easily recognized, and have good biostratigraphic value as well.

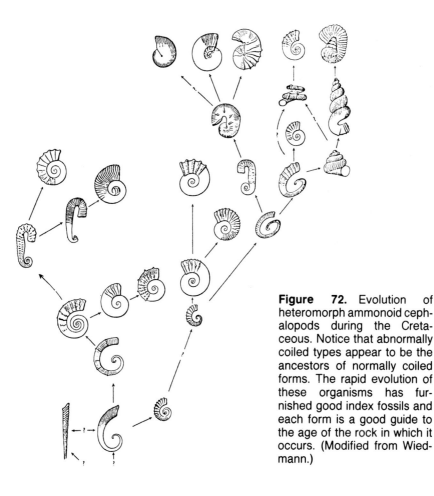

Figure 72. Evolution of heteromorph ammonoid cephalopods during the Cretaceous. Notice that abnormally coiled types appear to be the ancestors of normally coiled forms. The rapid evolution of these organisms has furnished good index fossils and each form is a good guide to the age of the rock in which it occurs. (Modified from Wiedmann.)

Evolution of Behavioral Patterns

Tracks and trails of organisms reflect the organism's structure, as was pointed out in the previous chapter. As the organism's structures evolve the tracks and trails reflect the changes. With this reasoning as a basis, it has been possible to use the evolution of traces as index fossils in biostratigraphy.

Seilacher (1970) has shown that in the early Paleozoic, *Cruziana*, or digging tracks (probably trilobite) changed from bilobed structures with scratch marks almost at right angles to the midline during the Cambrian to scratch marks that are in bundles almost parallel to the midline in younger Ordovician rocks (Fig. 73).

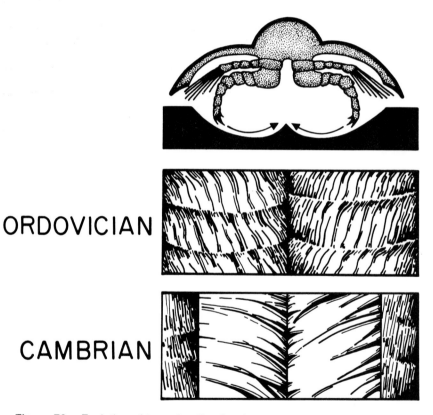

ORDOVICIAN

CAMBRIAN

Figure 73.   Evolution of trace fossils. *Cruziana,* or trilobite digging tracks, changed from almost perpendicular scratches in Cambrian to more or less parallel bundles of scratches in Ordovician. The change in scratch marks was related to evolving structure or behavior of the trilobites. The traces are excellent index fossils for the time indicated. (Adapted from original material and from several drawings by Seilacher.)

Thus, trace fossils illustrate evolutionary patterns that are useful in biostratigraphy. Of equal importance in the case of trilobite tracks is the fact that the traces show the evolution of structures that are not preserved on the trilobite body fossils. The traces indicate that the structure of the limbs steadily became better adapted to pushing the animal through the sediment in slightly different ways.

REFERENCES

Clark, D. L. 1965. Heteromorph ammonoids from the Albian and Cenomanian of Texas and adjacent areas. *Geological Society of America Memoir 95.*

Moore, R. C., ed. 1957. Part L—Mollusca 4, Cephalopoda, Ammonoidea. In *Treatise on invertebrate paleontology.* Geological Society of America and University of Kansas Press, Lawrence.

Seilacher, A. 1970. *Cruziana* stratigraphy of "non-fossiliferous" Palaeozoic sandstones. In *Trace fossils,* eds. T. P. Crimes, and J. C. Harper. Liverpool: Seel House Press.

Sweet, W. C., and Bergstrom, S. M., eds. 1971. Symposium on conodont biostratigraphy. *Geological Society of American Memoir 127.*

Wiedmann, J. 1969. The Heteromorphs and ammonoid extinction. *Biological Review* 44:563–602.

Chapter Eight

Fossils in Paleoecology

There are a number of ways in which fossils may aid in interpreting environments of the past. Some are direct, such as the observation that since echinoderms are presently all marine, fossil enchinoderms most likely are associated with marine sediment. Others are less direct, such as the observation that the ratio of certain oxygen isotopes in carbonate shells is a reflection of the oxygen content of the water in which the shell formed—and this, in turn, may be interpreted in terms of the water temperature. Most sound paleoecologic interpretations are based on careful study of fossils and their precise relationship to the enclosing sediment. Sediment characteristics as well as all of the factors of preservation, orientation, distribution, and frequency of the fossils are important. All of this information is more easily understood if the ecologic factors that control the distribution of modern organisms is understood. Indeed, without a knowledge of modern ecology, interpretations of paleoecology would be very difficult (Berry and Boucot, 1973). A few examples will illustrate this dependency.

WATER DEPTH AND PRESSURE

We have discussed already the fact that fossil cephalopods are among the most valuable organisms in biostratigraphy. Yet, living representatives of the important Paleozoic and Mesozoic groups have a considerable number of differences from their ancestors. Some shell types of the Paleozoic and Mesozoic (e.g., ammonoids) have no real living counterpart and their soft-part anatomy and physiology may have been equally different. How do we determine paleoecology?

The paleoecologist must assume that certain physical laws are more or less immutable. He tries to discover how these laws might be used to interpret certain structural characteristics of fossils. For example, cephalopod shells are characterized by internal chambering, the walls of which take the form of thin-walled spheres. The mechanical strength of the walls can be calculated and this applied to such things as the water depth in which the animals lived. Tensile strength of a spherical wall is directly proportional to the ratio of its thickness and the radius of curvature. Using this demonstrated physical law, Westermann (1972) measured the thickness and radius of curvature of fossil cephalopod walls in order to determine the greatest depth that these shells could have tolerated without imploding. Using the formula

$$\text{Stress} = \frac{\text{wall thickness} \times \text{internal pressure}}{\text{radius of curvature of wall}}$$

Westermann calculated the maximum hydrostatic pressure (depth) that the various fossil shells could tolerate. Then, by calibrating the actual depth range versus the theoretical range of modern forms against his calculation for fossil shells, Westermann was able to obtain depth ranges for twelve groups of cephalopods, ten of which have been extinct for millions of years. Thus, knowledge of physical laws can lead to paleo-ecologic interpretations.

TEMPERATURE

A similar kind of paleoecologic interpretation has been demon-strated for the water temperature in which certain planktonic Forami-nifera lived. *Globorotalia pachyderma* is abundant in the polar regions of the modern oceans. Bandy (1960) discovered that above 70° latitude, 95 to 100 percent of the *G. pachyderma* population coil to the left, evidently as an adaptation to colder water, while in the mid-latitudes, 95 to 100 percent coil to the right (Fig. 74). Thus, coiling direction in

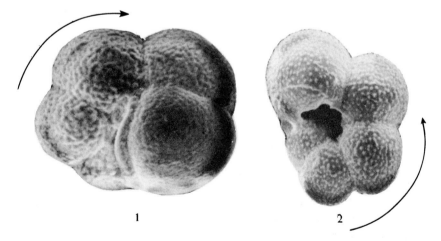

Figure 74. Coiling direction in Foraminifera. *Globorotalia* 1. sinistral coiling (cold water); 2. dextral coiling (warmer water). Specimens from the Arctic Ocean (X200).

this species appears to be temperature related. Because the Earth has experienced much colder temperatures (and continental glaciation) during the past two million years than earlier, the shifting of colder water masses north and south corresponding to colder climates can be determined from a study of the coiling direction of *G. pachyderma* in deep sea sediment. These studies have shown that the temperature of the water off the coast of southern California 10,000 years ago was more like modern North Atlantic and North Pacific temperatures. Further, these Foraminifera have furnished evidence that the Arctic Ocean probably has been frozen since its formation as a more or less isolated part of the Atlantic Ocean, 20 to 30 million years ago.

NUTRIENT LEVELS, FOOD CHAINS, AND OTHER ECOLOGIC FACTORS

The temperatures and depths of the oceans may be the paleo-ecologic factors most easily determined, but other ecologic factors are also obtainable from fossils. Nutrient levels, salinity and oxygen gradients from shorelines to deeper basins, water energy levels, and even fossil "hurricanes" can be interpreted for oceans 500 million years old.

How can a fossil shell, with no soft parts preserved, give information on the food chain in the oceans millions of years ago? The answer comes, as do all answers in paleontology, from careful observation and patient study. One example illustrates this. At the top of the modern marine food chain are the one celled plants and animals, and some of these have shells that can be preserved. One celled animals may eat one celled plants. Thus, modern Foraminifera feed on diatoms (one celled plant-like organisms). Cretaceous Foraminifera have been studied that show a consistent crop of diatoms in their shells. This then would be the top of the Cretaceous food chain. Cephalopods of the Cretaceous have been found with well-preserved Foraminifera in their body chamber—the next step in the chain. Finally, fossil cephalopod shells are very commonly found in the "bread basket" area of certain large marine reptiles of the Cretaceous — the end of the chain. (In fact, one study by Pollard (1968) described the fossilized stomach content of an ichthyosaur which contained fragments of at least 2500 different cephalopods. This compares favorably with modern sperm whales that have been found with 14,000 squid beaks that have accumulated in their intestinal track.) Participants in this marine food chain have changed, but the same pattern seen in modern oceans has been documented for the marine environment 100 million years ago.

Even the most minute details of the food chain have not escaped the eyes of the careful investigator. Recently Kaufman and Kesling (1960) reconstructed the fantastic pursuit of a cephalopod by a pursuing swimming reptile more than 100 million years ago. Only the fossil shell of the cephalopod has been found but its study shows that the reptile was able to sink its teeth into the shell of the cephalopod sixteen differ-ent times during the chase. From the study of the tooth points, the investigators were able to tell the direction and effectiveness of the various bites as well as the identity of the species of swimming reptile which was doing the biting. Evidently, the cephalopod was a good match for the reptile, at least for a while. The overlapping toothprints on the cephalopod shell were interpreted in their chronologic order indicating a final and sixteenth bite ended the chase. This final bite severed the entire body chamber of the cephalopod with all of the vital soft parts. From these observations, information was also gained as to the diet of this swimming reptile.

By way of summary, it should be emphasized that fossils must first be collected, studied and described. This still is the principal contribu-tion of paleontology. The knowledge of patterns of growth and develop-ment within populations and between communities can only be determined from good descriptions. From this, evolutionary develop-ment is often easily discernible. The second step is age determination for biostratigraphic and biogeographic studies, and then ecologic study for paleoclimatologic and paleoecologic information. The amount of data that is obtainable is greater now than ever before because of increased use of the tools of modern chemistry and physics. Ecologic and biologic data from the associations of organisms and determinations of the physical environment add more data. With so much data avail-able, the use of modern computers for analysis and correlation has found wide use among paleontologists. Thousands of factors—biologic, ecologic and geologic—can be analyzed in a very few minutes by use of large computers and with the many statistical methods now available.

REFERENCES

Bandy, O. L. 1960. The geologic significance of coiling ratios in the foraminifer *Globigerina pachyderma* (Ehrenberg). *Journal of Paleontology* 34:671–81.

Berry, W. B. N., and Boucot, A. J. 1973. Glacio-eustatic control of Late Ordovi-cian-Early Silurian platform sedimentation and faunal changes. *Geological Society of America Bulletin* 84:275–84.

Chamberlain, C. K., and Clark, D. L. 1973. Trace fossils and conodonts as evidence for deepwater deposits in the Oquirrh Basin of Central Utah. *Journal of Paleontology* 47:663–82.

Kaufman, E. G., and Kesling, R. V. 1960. An Upper Cretaceous ammonite bitten by a Mosasaur. *Contributions to the University of Michigan Museum of Paleontology* 15:193–248.

Pollard, J. E. 1968. The gastric contents of an Ichthyosaur from the Lower Lias of Lyme Regis, Dorset. *Palaeontology* 11:376–88.

Valentine, J. W. 1973. *Evolutionary paleoecology of the marine biosphere.* Englewood Cliffs, N.J.: Prentice-Hall.

Westermann, G. E. G. 1972. Strength of concave septa and depth limits of fossil cephalopods. *Lethaia* 6:383–403.

Chapter Nine

Fossils in Paleogeophysics

A few years ago a group of geologists and geophysicists gathered at Newcastle to consider the possibility that such things as the Earth's force of gravity, its magnetic field and its relationship to the Moon, might be different today than they were millions of years ago. More important was the idea that these paleophysical properties of the primitive Earth might be determined. These are physical properties easily determined today but how could such things have left any geologic record? The answers were to be found in the fossil record.

Fossils have proved extremely valuable to those interested in the physical properties of the Earth and in its history. For example, as was discussed in the previous chapter, fossils can be used for paleotemperature determinations. A second physical property that can be determined from fossils is that of the relationship of the Earth and Moon and the length and number of days during the geologic past. Another property that might be determined from fossils is the Earth's force of gravity during the past. Ancient tidal forces, or paleotides, constitute yet another possible area of study.

PALEOGEOPHYSICAL MEASUREMENTS

In the previous chapter, we discussed how certain fossils have been useful in determining the ocean temperatures of the Earth during the geologic past and the relationship of those temperatures to paleoclimatology. That fossils can be used as paleothermometers perhaps is less surprising than the fact that they can be used to determine the spatial relationship of the Earth and the Moon, the length of the day and year millions of years ago, and indirectly, as evidence that tidal forces were much stronger in the past than at present.

Fossils as tools in paleorotational studies for the Earth are best known. Wells (1963, 1970) and others have demonstrated that certain fossils have growth layers that correspond to the solar day, the synodical month, and, perhaps, the tropical year. For example, in certain corals there are individual growth bands or layers that correspond to a daily rhythm that in turn are a reflection of the Earth's daily temperature variation, or tidal flux, or nutrient supply. Modern corals show approximately the same number of bands as the number of days in our year. Wells reasoned that the wrinkled surface of a Paleozoic (Silurian) coral (Fig. 75) should reflect the same biological mechanism related in the same way to the Earth's rotation. He found that Paleozoic corals have a greater number of growth layers than modern corals and that there has been a progressive change through time. Thus, he discovered that an Ordovician coral from Ohio has 412 daily bands per year, a Middle Silurian coral from Sweden has more than 400, Devonian corals have from 385 to 405, Pennsylvanian forms have only 380, and so forth. A plot of the occurrence of these daily bands indicates the decrease in number of days since the Early Paleozoic (Fig. 76). It is interesting that species of molluscs as well as certain other extinct organisms give confirming evidence of a decrease in the number of days in the year since the Paleozoic. These observations clearly indicate a changing rotation rate of the Earth. The best explanation for this is that the Moon was much closer to the Earth in the Late Precambrian and Early Paleozoic than at present and that the gradual retreat of the moon from the Earth has resulted in a slower rotation rate. Fewer rotations mean fewer days.

One consequence of this interpretation is a better understanding of paleotides. The strength of the Earth's tides is inversely proportional to the cube of the Earth-Moon distance. Thus, if at one time the Moon had been closer to the Earth, tides would have been greater. Fossils and

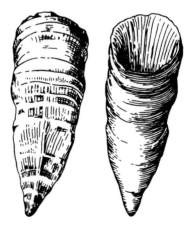

Figure 75. A rugose coral showing banded surface used to calculate length of year and paleorotation of Earth. Silurian (X1). (From Bayer et al.)

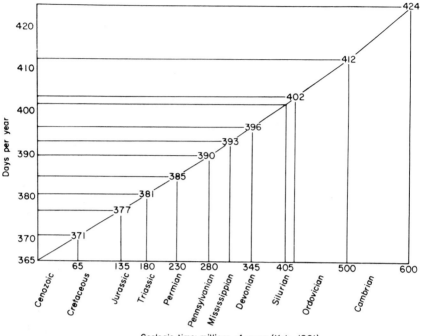

Geologic time: millions of years (Kulp, 1961)

Figure 76. Length of year during the geologic past based on number of growth bands in fossil corals. (From Wells.)

sediments should demonstrate this. Merifield and Lamar (1970) suggested that the strongly cross-bedded Late Precambrian and Cambrian sediments do demonstrate this fact. In addition, some observers have suggested that the rather rapid origin of shells as protective structures in invertebrates about 600 million years ago may be a reflection of Late Precambrian closeness of the Moon.

More speculative but of equal interest is the use of fossils for the interpretation of the strength of the Earth's gravity field through time. The gravitational force of the Earth is measured by

$$\text{Force} = \frac{g m_1 \times m_2}{d^2}$$

where g equals a gravitational factor; and m equals mass of the Earth and Moon; and d equals the distance between. This is for the present. How can it be measured for the past? Stewart (1970) has shown that such things as the size of flying animals, the size of large fossil vertebrates, and the depth of fossil footprints or traces are possible approaches to paleogravity. Gray (1968) has shown that the power output required for level flight in animals increases in direct proportion to the force of gravity. Mathematically, it can be demonstrated that the power necessary for level flight increases with size (and weight) more rapidly than flying animals are capable of producing and, thus, an upper limit for flight can be calculated. The fact that certain Cretaceous flying reptiles weighed up to 12 kg, and had a wing load of 2.58 kgm^{-2} (below the value for present birds of similar weight) suggests a significant change in the Earth's gravity field. The recent discovery of a Cretaceous flying reptile with a wingspan of 51 feet (the largest flying organism known) supports this idea. Also, many morphologists are convinced that the largest dinosaur could not support its mass of 80 metric tons in the present gravity field. Thus, the size of *Brontosaurus,* the largest tetrapod that has ever lived, can best be explained in terms of a relatively weak gravity field during the Cretaceous.

It has been suggested that perhaps the depth of footprints for dinosaurs could be studied in terms of pressure necessary to deform soft sediment as another approach to paleogravity (Stewart, 1970). Although none of the gravity studies are conclusive by themselves, it is apparent that fossils may have value in determining the strength of the Earth's gravity field in the geologic past. The preliminary studies summarized here suggest that 100 million years ago it was weaker than at present.

SUMMARY

The uses of fossils are many. They provide an insight into the function and behavior of life millions of years ago. Their evolution and distribution in rocks of the Earth have given man a relative time scale and a powerful tool for correlating rocks (and hence Earth events) for all parts of the Earth. Paleoclimatology, paleoecology, and paleogeophysics, have as their basis fossils that have recorded certain physical properties of a changing Earth.

REFERENCES

Gray, J. 1968. *Animal locomotion* London: Weidenfeld and Nicolson.
Merifield, P. M., and Lamar, D. L. 1970. Paleotides and the geologic record. In *Paleogeophysics,* ed. S. K. Runcorn. London: Academic Press.
Rosenberg, G. D. and Runcorn, S. K., eds. 1975. *Growth rhythms and the history of the Earth's rotation.* London: John Wiley.
Stewart, A. D. 1970. *Palaeogravity.* In *Paleogeophysics,* ed. S. K. Runcorn. London: Academic Press.
Wells, J. W. 1963. Coral growth and geochronometry. *Nature* 197:948–50.
Wells, J. W. 1970. Problems of annual and daily growth-rings in corals. In *Paleogeophysics,* ed. S. K. Runcorn. London: Academic Press.

Chapter Ten

Paleontology and Life Extraterrestrial

Each form of life which has occupied space on Earth has been unique. The Ordovician sponge-like animal with two walls separated by pillars is unlike any living thing today. A Cretaceous cephalopod with its large cane shaped shell as well as the Jurassic reptile *Allosaurus,* a fantastic reptilian carnivore, were all unique, but were representatives of "normal" life for the period of time in which they lived. Today's spider monkey or salmon would appear as curious to a Silurian-age observer as past forms of life may seem to us. And what is "normal" to our current view would seem odd to the observer two million years from now. It follows, that the final story of life of the Earth will not be complete even when all the knowledge gained from paleontology is combined with the knowledge of our modern fauna and flora. For living organisms continue to change, and the curious life forms of the year 3,750,000 A.D. will constitute but another chapter in the story of life. The life of any period of time of Earth's History is part of a continuously evolving lineage, and the most unusual form of life is not really so unusual in relationship to the events and time of its duration.

The time of a species' duration on Earth is important for our understanding of life of the past, present and future. The particular environmental and genetical factors which are responsible for a certain form of life do not remain long unchanged. The story of the Earth is the story of change. On land and in the water, temperatures, food supplies, and all of the complex ecologic factors upon which life is dependent, are ephemeral, and it is a particular environment and a particular genetical system which interact to produce each form of life. The appearance of a certain type of organism during a specific period of geologic time was due to the fact that all of the internal and external factors which are collectively responsible, interacted in a particular way during that time interval. If any of these factors had been different, or had any worked together in a different way or at a different time, a different organism would probably have been produced. Thus, the giant among swimming reptiles, *Plesiosaurus,* would not have been a *Plesiosaurus* under any other set of external and internal interactions except those which led to its development. Higher organisms have 100^{1000} possible number of gene combinations and the number of possible ecologic factors is also extremely large. If these external and internal interactions are constantly changing, it appears unlikely that two species of exactly identical biologic makeup would ever develop, especially when separated by millions of years of change.

Herein is the explanation of each "unique" form of life. And herein too, rests the basis for speculation concerning extraterrestrial life.

Why this speculation at all? In spite of all of the meteorites that have reached the earth with their cargoes of supposedly "organic" material, none studied to date has been widely accepted as having been formed by life processes as we understand them. Currently, there is little direct evidence for life except on this Earth. Current planetary studies may, of course, change this notion, and, in any case, it would be a sign of folly to maintain dogmatically that there is no life extraterrestrially. The next meteorite to arrive may contain definite organically formed matter. Further, the lack of positive evidence concerning the existence of life elsewhere in the universe does not rule out its possibility. In fact, there seems to be widespread belief that there is much life elsewhere, and the evidence, although indirect, is the subject of much current interest.

Although some of this interest is reflected on the comic page where Martian monsters with three eyes perform, more sober heads ponder a serious approach to this study. A unique environment and events just statistically probable have set in motion several billion years of evolution on the Earth. The life produced by these unusual circumstances is everywhere around us and as part of it, we now can contemplate if, and how, a similar chain of events may have been set in motion elsewhere in our universe. While we may give the comics a quick Sunday-morning scanning, the ideas now being formed about life in our universe merit more than such a perfunctory glance.

We must admit, at first, that we know of no other planet where conditions exactly like those on Earth have ever existed. Such ignorance does not rule out the possibility, however.

There are apparently at least two opposite views on life extraterrestrial. The one maintains that there *must* be life present elsewhere and that this life may be at the same or at least in a similar stage of evolution as we believe ours is. Whether this is in our solar system or some other is not important. The other view is that because there is no evidence for life elsewhere, it is useless to speculate on its existence, and further, even if life does exist elsewhere in this or some other solar system, by its very nature, it would be considerably different from life on the Earth.

There are various champions of these different views but, in general, non-biologically minded students maintain that life comparable to that on Earth exists elsewhere, while the more biologically minded students strongly maintain that if life does exist elsewhere, it would most likely be quite different.

These different ideas are based on a different understanding (or lack of understanding) of the processes responsible for life. Life is being progressively reduced from a vitalistic to a chemical process. At the present time, life is literally in the hands of the chemist, the biochemist in particular. However, while the actions, reactions, and interactions of DNA, RNA, and other chemical structures may explain origin and physiology, it would seem apparent to most observers that the biologist, who is concerned with the evolution and development of life, would have greater insight into the nature of the kind of life which would ultimately appear under given conditions.

The possibility that life elsewhere may be like that on Earth is supported by several ideas. One is that life from Earth may have been transported to other places in the solar system long before our first rockets performed this service. For instance, two of the principal theories for the origin of the Moon support the idea that the Moon was either derived from the Earth directly or that it was derived from another celestial body by collision with and then capture by the Earth. In either of these theories, it could be supposed that the event occurred when there was some form of primitive life on the Earth and that from either derivation, the accompanying fragments took some of this life with it. Because many of the meteorites with organic-like material may have ultimately been related to such an event it is also likely that other planets may have received transplants from mother earth in the same way. With such a common origin, why couldn't the product of evolution, several billion years later, be approximately the same?

A different theory, based on assumed similarities in environmental conditions, has been used by yet other students of this problem. In support of this theory, the number of star-planetary systems possible in our universe is calculated. Further calculations of masses and other apsects of these systems then lead to the conclusion that there may be as many as four planets per visible star and that under the conditions of the calculations, approximately two planets per visible star may be in the "heat flux" zone, the space interval which could support life as we know it on Earth. From these figures, it would follow that in our galaxy alone, some 10^{11} planetary systems may be available with potential Earth-like environments. The conclusion is that "man is not alone," and that his equivalent could be on hundreds or thousands of planets. It would seem to be a long way from the "heat flux" zone similarities to man's equivalent, but a natural corollary to the assumption of these similarities, according to the supporters of this theory, is that forms of life equivalent to all Earth's plants and animals may be abundant only a hop-skip and a light year away.

More biologically oriented students would take issue at this point. Although one may accept the calculations for the abundance of Earth-like planets which *may* support life, our knowledge of the processes involved in evolution, at least on this planet, would suggest that any life present elsewhere, would, of necessity, be considerably different from that which has developed on Earth. Given even the same physical and chemical factors in a gross sort of way and the same potential for development, there would still be variables of tremendous magnitude. The timing of the interactions of the varying micro-environmental factors would not be the same. There are many other unknowns. Under different gravity, magnetic, or other physical forces, how do chemical and biologic systems interact? Especially over a long period of time it would seem apparent that considerable differences in any resulting forms of life are at least probable. Our current space experiments may enlighten us here, but the long-term effects may only be apparent from long-term observation.

Because of the uniqueness of environment, genetics, and time factors, it appears unlikely that man's equivalents, at least in a recognizable form, could exist elsewhere, at least under the conditions proposed by the proponents of the theory described above.

Homeomorphs, or very similar organisms, unrelated in origin, and separated by millions of years, do occur in the paleontologic record. These homeomorphs are never identical, however, and even when produced in strikingly similar environments there are differences of great magnitude.

Does any form of life exist extraterrestrially? The evidence derived from meteorites and the theoretical calculations for environmental conditions in our galaxy seems to be positive. On the other hand, our understanding of the chemical and biologic factors of evolution suggests that whatever kind of life it may be, it would probably show considerable difference from that which exists today or has existed for several billion years on Earth.

If our sun holds out, there may be another 5 billion years of evolution. Given this possibility, anything can happen.

REFERENCES

Kaplan, S. A., ed. 1969. *Extraterrestrial civilizations: problems of interstellar communication.* Moscow (in Russian).
Klein, H. P. 1972. Potential targets in the search for extraterrestrial life. In *Exobiology,* ed. C. Ponnamperuma. Amsterdam: North-Holland Publishers.

Sagan, C. 1972. Life beyond the solar system. In *Exobiology,* ed. C. Ponnam-
    peruma. Amsterdam: North-Holland Publishers.
Shklovskii, I. S., and Sagan, C. 1966. *Intelligent life in the universe.* San Fran-
    cisco: Holden-Day.

Index